CW01460912

NEUROSCIENCE RESEARCH PROGRESS

VISUAL CORTEX: ANATOMY, FUNCTIONS AND INJURIES

NEUROSCIENCE RESEARCH PROGRESS

Additional books in this series can be found on Nova's website under the Series tab.

Additional E-books in this series can be found on Nova's website under the E-book tab.

NEUROLOGY - LABORATORY AND CLINICAL RESEARCH DEVELOPMENTS

Additional books in this series can be found on Nova's website under the Series tab.

Additional E-books in this series can be found on Nova's website under the E-book tab.

NEUROSCIENCE RESEARCH PROGRESS

VISUAL CORTEX: ANATOMY, FUNCTIONS AND INJURIES

JESSICA M. HARRIS
AND
JASON SCOTT
EDITORS

Nova Science Publishers, Inc.
New York

Copyright © 2012 by Nova Science Publishers, Inc.

All rights reserved. No part of this book may be reproduced, stored in a retrieval system or transmitted in any form or by any means: electronic, electrostatic, magnetic, tape, mechanical photocopying, recording or otherwise without the written permission of the Publisher.

For permission to use material from this book please contact us:
Telephone 631-231-7269; Fax 631-231-8175
Web Site: http://www.novapublishers.com

NOTICE TO THE READER

The Publisher has taken reasonable care in the preparation of this book, but makes no expressed or implied warranty of any kind and assumes no responsibility for any errors or omissions. No liability is assumed for incidental or consequential damages in connection with or arising out of information contained in this book. The Publisher shall not be liable for any special, consequential, or exemplary damages resulting, in whole or in part, from the readers' use of, or reliance upon, this material. Any parts of this book based on government reports are so indicated and copyright is claimed for those parts to the extent applicable to compilations of such works.

Independent verification should be sought for any data, advice or recommendations contained in this book. In addition, no responsibility is assumed by the publisher for any injury and/or damage to persons or property arising from any methods, products, instructions, ideas or otherwise contained in this publication.

This publication is designed to provide accurate and authoritative information with regard to the subject matter covered herein. It is sold with the clear understanding that the Publisher is not engaged in rendering legal or any other professional services. If legal or any other expert assistance is required, the services of a competent person should be sought. FROM A DECLARATION OF PARTICIPANTS JOINTLY ADOPTED BY A COMMITTEE OF THE AMERICAN BAR ASSOCIATION AND A COMMITTEE OF PUBLISHERS.

Additional color graphics may be available in the e-book version of this book.

LIBRARY OF CONGRESS CATALOGING-IN-PUBLICATION DATA

Visual cortex : anatomy, functions, and injuries / editors, Jessica M. Harris and Jason Scott.
p. ; cm.
Includes bibliographical references and index.
ISBN 978-1-62100-948-1 (hardcover)
I. Harris, Jessica M. II. Scott, Jason.
[DNLM: 1. Visual Cortex. WL 307]

548'.842--dc23

2011040545

Published by Nova Science Publishers, Inc. † New York

CONTENTS

PREFACE

The visual thalamus conveys visual information detected by the retina to the visual cortex along parallel pathways with distinct anatomical and physiological characteristics. This group of pathways is comprised of the magnocellular, the parvocellular, and the koniocellular pathways. In this book, the authors present research on the anatomy, functions and injuries of the visual cortex. Topics include the biophysical mechanisms underlying the critical period of visual cortical plasticity; electrophysiological assessment of the human visual system; plasticity of visual cortical circuitries in adulthood; the visual cortex in Alzheimer's disease and molecular signatures of parallel pathways in the visual thalamus.

Chapter 1 – Neural circuits are shaped by reflecting the sensory experience in the restricted critical period during postnatal life. Although the importance of this critical period plasticity is well understood, from a clinical and scientific viewpoint, the underlying molecular mechanisms are largely unclear. In the visual cortex, GABA has been suggested to play a role in triggering the onset of critical period for experience-dependent ocular dominance plasticity. However, another line of evidence suggests that the developmental switch in the NMDA receptor (NMDAR) subunit composition, from NR2B to NR2A, may be involved in controlling the timing of plasticity. In this study, to explore the integrative role of GABA and NMDARs in developmental visual cortical plasticity, I theoretically examine the synaptic dynamics by spike-timing-dependent plasticity (STDP) by extending a previously published cortical STDP model [Kubota and Kitajima, Journal of Computational Neuroscience, 28, 347-359, 2010] to take into account the intracellular Ca^{2+}-based mechanism of plasticity. I show by using a pairing protocol simulation that the increase in the peak NMDAR conductance and the prolongation in the NMDAR decay time strengthen LTP more significantly than LTD in STDP. Based on this finding, I propose a novel biophysical model of STDP, in which activity- and subunit-dependent desensitization of NMDARs dynamically modulates the LTP/LTD balance under background GABA inhibition. I demonstrate that when a neuron receives homogeneous random inputs, the LTP/LTD ratio is kept at around one and STDP exhibits strong regulatory function that maintains approximate excitatory-inhibitory balance. When a neuron receives two groups of correlated inputs, as in the case of visual cortical neurons receiving inputs from two eyes, competitive interaction occurs between the different input groups only in the presence of mature states of both NMDAR subunit composition and GABA inhibition. The competitive function is required for the synaptic pattern to reflect the past history of input activities and to provide the basis of experience-dependent synaptic modifications. These results may suggest that the GABA and NMDARs would cooperate to trigger cortical critical period and to embed early sensory experience into synaptic pattern.

Chapter 2 – In humans, visual information is processed simultaneously via multiple parallel channels. Condensed and parallel signals from the retina arrive in the primary visual cortex via the lateral geniculate nucleus. These signals then remain segregated until the higher levels of visual cortical processing through at least two separate but interacting parallel pathways; the ventral and dorsal streams. The former projects to the inferior temporal cortex for processing form and color, because it can detect visual stimuli with high spatial frequency and color. In contrast, the latter connects to the parietal cortex for detecting motion, because it responds to high temporal frequency stimuli. Based on these distinct physiological characteristics, we hypothesized that manipulating visual stimulus parameters would enable us to evaluate the different levels of each stream. So far, we have developed several techniques to record visual evoked potentials (VEPs) and event-related potentials (ERPs) with optimal stimuli. In this review, we first summarize current concepts of the major parallel visual pathways. Second, we describe the relationship between the parallel visual pathways and higher visual system dysfunction. Third, we introduce VEP and ERP techniques that can assess the function of each stream and region of visual cortex. Finally, we address the clinical applications of VEP and ERP recording techniques for several neurological disorders involving specific visual dysfunction.

Chapter 3 – The formation of neural circuitries in the brain relies on a tight interaction between genes and environment. As intrinsic factors mediate the initial assembly of synaptic circuitries, the nervous system begins to process sensory information thus creating neuronal representations of the external world that are continuously modified by experience. Sensory experience, in fact, modifies the structural and functional architecture of the nervous system in response to changing environmental conditions throughout life. This phenomenon is particularly evident during early stages of development when experience drives the consolidation of synaptic connectivity but the reorganization of neural circuitries continues in adulthood, as for instance, in response to learning, loss of sensory input or trauma. How does experience modify synaptic circuitries in the visual system? This will be one of the topics I shall concentrate on inthis chapter.

The notion that neural circuitries in the adult brain can change in response to experience has become a major conceptual subject in modern neuroscience.Althoughearly studies dealt with plasticity of the developing nervous system,the study of adult brain plasticity and the identification of molecular and cellular mechanisms at the basis of these plastic phenomena are current challenges in the neuroscience field. This chapter outlines physiological processes that underlie neuronal plasticity in the visual cortex and provides an in-depth discussion about the enhancement of plasticity as a strategy for brain repair in adulthood. Given that experience-dependent changes of brain functions depend, at least partially, on the expression of genes that have evolved to meet specific environmental demands, molecular mechanisms that lie behind processes of neuronal plasticity shall also be addressed. How intracellular signal transduction pathways associated to sensory experience regulate epigenetic modifications of chromatin structure that underlie visual cortical plasticity is also a novel concept that will be considered.

Chapter 4 – Alzheimer's disease (AD) is an important neurodegenerative disorder causing visual problems in the elderly population. The pathology of AD includes the deposition in the brain of abnormal aggregates of β-amyloid (Aβ) in the form of senile plaques (SP) and abnormally phosphorylated tau in the form of neurofibrillary tangles (NFT). A variety of visual problems have been reported in patients with AD including loss of visual acuity (VA), colour vision and visual fields; changes in pupillary responses to mydriatics, defects in fixation and in smooth and saccadic eye movements; changes in contrast sensitivity and in visual evoked potentials (VEP); and disturbances in complex visual tasks such as reading, visuospatial function, and in the naming and identification of objects. In addition, pathological changes have been observed to affect the eye, visual pathway, and visual cortex in AD. To better understand degeneration of the visual cortex in AD, the laminar distribution of the SP and NFT was studied in visual areas V1 and V2 in 18 cases of AD which varied in disease onset and duration. In area V1, the mean density of SP and NFT reached a maximum in lamina III and in laminae II and III respectively. In V2, mean SP density was maximal in laminae III and IV and NFT density in laminae II and III. The densities of SP in laminae I of V1 and NFT in lamina IV of V2 were negatively correlated with patient age. No significant correlations were observed in any cortical lamina between the density of NFT and disease onset or duration. However, in area V2, the densities of SP in lamina II and lamina V were negatively correlated with disease duration and disease onset respectively. In addition, there were several positive correlations between the densities of SP and NFT in V1 with those in area V2. The data suggest: (1) NFT pathology is greater in area V2 than V1, (2) laminae II/III of V1 and V2 are most affected by the pathology, (3) the formation of SP and NFT in V1 and V2 are interconnected, and (4) the pathology may spread between visual areas via the feed-forward short cortico-cortical connections.

Chapter 5 – Based on the probabilistic reasoning to vision information processing in this paper, combining synchronized response and sparse representation, we propose a new early visual computational model, which consists of the multi-scale filtering, the phase synchronization and the inner product operations. According to features of parallel distributed processing in the visual pathway, the retinal image may be orthogonally divided to sub-unit of image or image patch by means of size of receptive field of ganglion cell, then, all of the information contents contained in image patches transmitted by sub-channel to the primary visual cortex V1, respectively. Further processing to them are made by these distributed cortical functional columns, and its spatial localized, oriented and band-pass characteristics can made response to features of image patches, and from this the realization of inner product operations is achieved, it is an optimal detection operator under the meaning of minimum mean square error to visual image reconstruction. Theoretical analysis and experimental results showed that: at the system level inner product operator reflects the nature of the excitation process of local characteristics of external stimuli onto corresponding neurons in the cortex V1. It is also a plausible assumption about neural computation in V1 cortex. Therefore, it may be have some reference significance to explore the neural mechanisms of the visual information processing. In addition, difficulties, which occur in simulating the receptive field of simple cell in the V1 cortex by sparse coding, are briefly discussed, and the problems, which arise when the cortical functional column is considered as a tiled set of selective filter, also will be compared and analyzed.

Chapter 6 – The activity-regulated cytoskeleton-associated protein gene *(Arc)* is one of the immediate early gene markers in the visual cortex up-regulated by light stimuli. The expression of *Arc* in the brain correlates with various sensory processes, natural behaviors, and pathological conditions. Arc is also involved in synaptic plasticity during development. Thus, *in vivo* monitoring of *Arc* expression is useful for the analysis of physiological and pathological conditions in the brain. We generated a novel transgenic mouse strain to monitor the neuronal-activity-dependent *Arc* expression using bioluminescence signals *in vivo*. Using the bacterial artificial chromosome (BAC) containing the entire mouse *Arc*, we introduced the firefly-derived luciferase (Luc) gene at the translational initiation site of *Arc* by homologous recombination in *Escherichia coli (E. coli.)* to generate the BAC transgene construct. We injected the construct into fertilized one-cell embryos and obtained transgenic mouse strains. Immunohistochemical analysis revealed the strong coexpression of endogenous Arc and exogenous Luc in the neuronal soma in layers 4 and 6 in the visual cortex. Because of the very high sensitivity with a high signal-to-noise ratio, we successfully detected the changes in bioluminescence signal intensity in the visual cortex under the light and dark conditions. These changes correlated well with the changes in the expression levels of Arc and Luc examined by Western blot analysis. Visual deprivation by monocular eye enucleation (ME) resulted in the significant decrease in bioluminescence signal intensity in the contralateral posterior region within 24 hr. Interestingly, one month after ME, there was no significant difference in bioluminescence signal intensity between the right and left visual areas. These neuronal-activity-dependent plastic changes in the bioluminescence signal intensity in the mouse visual cortex after visual deprivation suggest structural plasticity after peripheral lesions in adults. Our novel mouse strain will be valuable for the continuous monitoring of neuronal-activity-dependent *Arc* expression in the visual cortex under physiological and pathological conditions.

Chapter 7 – The visual thalamus conveys visual information detected by the retina to the visual cortex along parallel pathways with distinct anatomical and physiological characteristics. This group of pathways is comprised of the magnocellular, the parvocellular, and the koniocellular pathways. Although considerable progress has been made with regard to our knowledge of the anatomical circuitry and physiological properties that distinguish these three parallel pathways, our molecular understanding of the parallel pathways is still insufficient. This is, at least partially, because these pathways are not well-developed in mice, which are commonly used for molecular investigations. Recent studies of these pathways, therefore, have employed higher mammals such as primates and carnivores. In this chapter, I summarize recent findings regarding the molecular investigations of the visual thalamus, especially focusing on the parallel pathways in higher mammals. I also discuss issues to be addressed and potential future directions of this field.

In: Visual Cortex: Anatomy, Functions … ISBN: 978-1-62100-948-1
Editors: J.M. Harris et al. pp. 1-35 © 2012 Nova Science Publishers, Inc.

Chapter 1

BIOPHYSICAL MECHANISMS UNDERLYING THE CRITICAL PERIOD OF VISUAL CORTICAL PLASTICITY: A MODELING STUDY

Shigeru Kubota
Graduate School of Science and Engineering,
Yamagata University, 4-3-16 Jonan,
Yonezawa, Yamagata, 992-8510, Japan,
Tel: +81-238-26-3585; Fax: +81-238-26-3240,
E-mail: kubota@yz.yamagata-u.ac.jp

ABSTRACT

Neural circuits are shaped by reflecting the sensory experience in the restricted critical period during postnatal life. Although the importance of this critical period plasticity is well understood, from a clinical and scientific viewpoint, the underlying molecular mechanisms are largely unclear. In the visual cortex, GABA has been suggested to play a role in triggering the onset of critical period for experience-dependent ocular dominance plasticity. However, another line of evidence suggests that the developmental switch in the NMDA receptor (NMDAR) subunit composition, from NR2B to NR2A, may be involved in controlling the timing of plasticity. In this study, to explore the integrative role of GABA and NMDARs in developmental visual cortical plasticity, I theoretically examine the synaptic dynamics by spike-timing-dependent plasticity (STDP) by extending a previously published cortical STDP model [Kubota and Kitajima, Journal of Computational Neuroscience, 28, 347-359, 2010] to take into account the intracellular Ca^{2+}-based mechanism of plasticity. I show by using a pairing protocol simulation that the increase in the peak NMDAR conductance and the prolongation in the NMDAR decay time strengthen LTP more significantly than LTD in STDP. Based on this finding, I propose a novel biophysical model of STDP, in which activity- and subunit-dependent desensitization of NMDARs dynamically modulates the LTP/LTD balance under background GABA inhibition. I demonstrate that when a neuron receives homogeneous random inputs, the LTP/LTD ratio is kept at around one and STDP exhibits strong regulatory function that maintains approximate excitatory-inhibitory balance. When a neuron receives two groups of correlated inputs, as in the case of visual cortical neurons receiving inputs from two eyes, competitive interaction occurs between the different input groups only in the presence of mature states of both NMDAR subunit composition and GABA inhibition. The competitive function is required for the synaptic pattern to reflect the past history of input activities and to provide the basis of experience-dependent synaptic modifications. These results may suggest that the GABA and NMDARs would cooperate to trigger cortical critical period and to embed early sensory experience into synaptic pattern.

1. Introduction

The brain has a high ability of plasticity during postnatal critical period, which permits sensory experience in early development to be embedded into neural connections (Wiesel 1982). A representative example is the loss of responsiveness in the visual cortical neurons to an eye that has been briefly deprived of visual stimuli. There are mainly two hypotheses regarding the molecular mechanisms underlying the critical period. An important candidate is the cortical GABAergic system which mediates major inhibitory synaptic transmission (Hensch 2005). Several experiments have shown by using genetic and/or pharmacologic manipulation of GABAergic transmission that the strengthening of intracortical GABA advances the critical period of ocular dominance plasticity in the visual cortex, whereas the suppression of GABA eliminates the ability of this type of plasticity until the inhibition level is recovered again (Hensch et al. 1998; Hanover et al. 1999; Huang et al. 1999; Fagiolini and Hensch 2000; Hensch 2005). These findings strongly suggest that the maturation of GABA can trigger critical period in the visual cortex *in vivo*, although the fact that GABA does not affect (Hensch et al. 1998; Froemke and Dan 2002) or suppresses (Kirkwood and Bear 1994) long-term plasticity in the visual cortex *in vitro* makes the underlying synaptic mechanism difficult to understand.

However, the GABA control hypothesis does not agree with another line of evidence that the developmental change in the subunit composition of NMDA receptors (NMDARs) may be involved in regulating the timing of critical period (Roberts and Ramoa 1999; Erisir and Harris 2003; Daw et al. 2007). NMDARs are tetramers composed of obligatory NR1 and variable NR2 (which includes NR2A and NR2B) subunits (Stephenson 2001), and the ratio of NR2A to NR2B subunits increases over development (Monyer et al. 1994; Flint et al. 1997; Quinlan et al. 1999a, b; Mierau et al. 2004). The alteration in the NMDAR subunit composition occurs at the same time period as the critical period for experience-dependent plasticity (Roberts and Ramoa 1999; Erisir and Harris 2003; Daw et al. 2007) as well as that for NMDAR-dependent long-term potentiation (LTP) and depression (LTD) of cortical synapses (Crair and Malenka 1995; Feldman et al. 1998). These findings suggest that the critical period may be controlled through the time course of NMDAR subunit gene expression.

Recently, I have proposed a computational model that examines an integrative role of NMDAR subunit expression and GABA inhibition in synaptic organization through spike-timing-dependent plasticity (STDP) (Kubota and Kitajima 2010). In this model, the activity-dependent feedback (ADFB) mechanism in STDP has been incorporated, in which the magnitude of LTP is decreased by the higher postsynaptic activity through the activity-dependent desensitization of NMDARs (Kubota et al. 2009; Nevian and Sakmann 2006; Caporale and Dan 2008). With the ADFB function, the LTP/LTD ratio in STDP is negatively regulated by the feedback of postsynaptic activity, producing a dynamic balance between LTP and LTD (Kubota et al. 2009; Kubota and Kitajima 2010). The strengthening of ADFB modulation of LTP, through the mature states of NMDAR subunit expression, as well as the enhancement of background GABA inhibition have been shown to be necessary for the induction of experience-dependent synaptic modifications (Kubota and Kitajima 2010).

However, it is still largely unclear how the cortical synapses can detect the postsynaptic activity and adequately modulate the LTP/LTD balance as proposed by the previous study (Kubota and Kitajima 2010). Furthermore, there is accumulating evidence that the activation of NMDARs and the resulting Ca^{2+} influx through NMDARs may be highly involved in the induction of both LTP and LTD in various forms of plasticity including STDP (Bi and Poo 1998; Debanne et al. 1998; Feldman 2000; Magee and Johnston 1997; Markram et al. 1997; Sjöström et al. 2001; Zhang et al. 1998). This implies that, different from the previous model, the activity-dependent desensitization of NMDARs could decrease both LTP and LTD, which would make it difficult to reliably control the direction of the change in the LTP/LTD ratio.

In the present study, I extend the previous study to take into account the role of spine Ca^{2+} signals in STDP (Nevian and Sakmann 2006). I perform a simulation of the pairing protocol using a biophysical Ca^{2+}-dependent plasticity (CaDP) model (Kubota and Kitajima, 2008), which agrees with a standard hypothesis of CaDP that strong Ca^{2+} elevation induces LTP whereas moderate and prolonged Ca^{2+} elevation elicits LTD (Yang et al. 1999; Mizuno et al. 2001; Taniike et al. 2008; Sjöström and Nelson 2002). I demonstrate that the LTP/LTD ratio in STDP can be controlled by NMDARs such that larger and more prolonged NMDAR activation leads to significant increase in this ratio. Based on this finding, I examine a novel model of STDP, in which the activity- and subunit-dependent desensitization of NMDARs dynamically modulates LTP/LTD balance according to the predicted outcome of the paring protocol simulation, in the presence of background GABA inhibition. I demonstrate that when a neuron receives random inputs, the LTP/LTD ratio is kept around one and STDP displays strong regulatory action maintaining an approximate balance between excitation and inhibition. Furthermore, when a neuron receives two groups of correlated inputs, as in the case of visual cortical cells driven by the inputs from both the eyes, the competitive interaction between the different input groups, necessary for experience-dependent synaptic modifications, emerges through the maturation of both NMDAR subunit expression and GABA inhibition. These findings may suggest that the CaDP hypothesis of plasticity is consistent with the idea that the LTP/LTD ratio in STDP is subject to metaplastic regulation through the feedback of postsynaptic activity (Kubota et al. 2009; Kubota and Kitajima 2010). In addition, the present study further supports the prediction of the previous study (Kubota and Kitajima 2010) that the NMDAR- and GABA-dependent mechanisms can cooperate to trigger the critical period through activity-dependent competition between afferent inputs.

2. METHODS

2.1. Neuron Model

We used a two-compartment conductance-based pyramidal neuron consisting of a soma and a dendrite (Kubota and Kitajima 2010) (Fig. 1A). The membrane potentials of the somatic and dendritic compartment, V_s and V_d (in mV), obey the following equations:

$$C_m \frac{dV_s}{dt} = -I_{leak} - I_{Na} - I_K + \frac{g_c}{p}(V_d - V_s) + I_{inj}, \tag{1}$$

$$C_m \frac{dV_d}{dt} = -I_{leak} - I_{Na} - I_K - I_{Ca,V} - I_{AHP} + \frac{g_c}{1-p}(V_s - V_d) - I_{syn}. \tag{2}$$

Here, C_m (= 1 μF/cm^2) is the specific membrane capacitance, I_{leak} is the leak current, I_{Na} and I_K are the voltage-gated sodium and potassium currents, respectively, $I_{Ca,V}$ is the high-threshold voltage-gated Ca^{2+} currents, and I_{AHP} is the Ca^{2+}-dependent K^+ currents. I_{inj} and I_{syn} are the injected current to the soma and the synaptic current to the dendrite, respectively. g_c (= 2 mS/cm^2) represents the coupling conductance between the two compartments and p (= 0.5) is the ratio of soma to total area (Wang 1998).

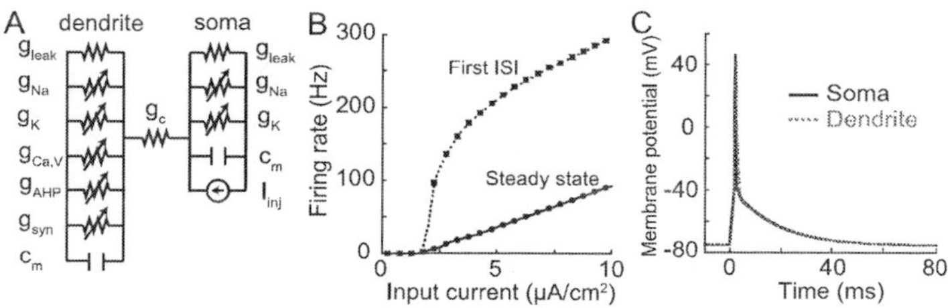

Figure 1. Conductance-based pyramidal neuron model. (A) The conductances incorporated in the somatic and dendritic compartments. (B) The f-I relationship of the model. The instantaneous spike frequency for the first spike interval and the steady-state firing frequency are shown by the dashed and solid lines, respectively. (C) The action potential obtained displays a waveform with a slow component of ADP.

The model neuron reproduces the initial and adapted *f-I* curves similar to those of neocortical pyramidal cells (Fig. 1B) (Ahmed et al. 1998). Furthermore, the action potential waveform of this model displays slow afterdepolarization (ADP) found in pyramidal cells (Fig. 1C) (Feldman 2000; McCormick et al. 1985). Slow ADP is required in our model to produce a moderate level of NMDAR activation, which is responsible for LTD induction when a presynaptic activation follows a postsynaptic spike (Kubota and Kitajima 2008; Shouval et al. 2002).

2.2. Synaptic Currents

For the synaptic inputs, we consider NMDAR- and AMPA receptor- (AMPAR-) mediated excitatory currents and GABA-mediated inhibitory currents (Kubota and Kitajima 2010). The AMPA conductance obeys the following alpha function

$$G_{AMPA}(t) = w\, g_{AMPA}\,(e / t_{AMPA})\, t \exp(-t / t_{AMPA}), \tag{3}$$

where w is the synaptic weight, g_{AMPA} = 2.5 μS/cm^2 is the peak conductance, and t_{AMPA} = 1.5 ms (Zador et al. 1990). The weight w is set to 1 in the simulation of the pairing protocol (Figs. 2 and 3), while its value is dynamically changed in the simulation of synaptic weight modification using random inputs (Figs. 4–9). The NMDAR conductance follows the equation dependent on the postsynaptic membrane voltage:

$$G_{NMDA}(t) = g_{NMDA} \frac{\exp(-t/\tau_{decay}^{NMDA}) - \exp(-t/\tau_{rise}^{NMDA})}{1 + 0.33\exp(-0.06V(t))}, \tag{4}$$

where g_{NMDA} (see below) denotes the peak conductance, τ_{decay}^{NMDA} (see below) and τ_{rise}^{NMDA} = 0.67 ms are the decay and rise time constants, respectively (Jahr and Stevens 1990; Koch 1999). The GABA conductance follows the following equation

$$G_{GABA}(t) = g_{GABA}\,(e / t_{GABA})\, t \exp(-t / t_{GABA}), \tag{5}$$

where g_{GABA} denotes the peak conductance and t_{GABA} = 10 ms (Bernander et al. 1991). The reversal potentials associated with these synaptic conductances are 0 mV and –70 mV for excitatory and inhibitory currents, respectively. The synaptic conductances evoked by the past inputs are summed linearly for calculating the values of the conductances at the present time.

2.3. Ca^{2+} Dynamics in the Spine

In the simulation for the pairing protocol (Figs. 2 and 3), we considered the Ca^{2+} dynamics in a dendritic spine. We simply assumed that the spine is electrically fully coupled to the dendrite but chemically isolated from it (Koch 1999; Koch and Poggio 1983; Sabatini et al. 2002); in other words, the spine and dendrite are isopotential, whereas their Ca^{2+} dynamics are independent of each other. In addition, we assumed that excitatory synapses lie on dendritic spines (Koch 1999). With this assumption, the Ca^{2+} dynamics in the spine is expressed as follows:

$$\frac{d[Ca]_{sp}}{dt} = -\frac{[Ca]_{sp}}{\tau_{sp}} - \alpha_{sp}(I_{Ca,V} + I_{Ca,NMDA}) \cdot \qquad (6)$$

In this equation, τ_{sp} = 20 ms and $\alpha_{sp} = A_{sp}/(2Fv_{sp})$, where F is the Faraday constant, the spine membrane area A_{sp} = 2.7 μm^2, and its volume v_{sp} = 0.29 μm^3 (Harris et al. 1992; Sabatini et al. 2002). The conductance associated with the voltage-gated calcium current $I_{Ca,V}$ (taken from Kubota and Kitajima (2010)) is $g_{Ca,V}$ = 0.03 mS/cm^2, which is determined such that the amplitude of the spike-induced Ca^{2+} increase in the spine (~1.6 μM) is physiologically relevant (Sabatini et al. 2002). The Ca^{2+} current through NMDARs, $I_{Ca,NMDA}$, is expressed as follows:

$$I_{Ca,NMDA} = G_{Ca,NMDA}(t)(V - E_{Ca}), \qquad (7)$$

$$G_{Ca,NMDA}(t) = k_{sp} g_{NMDA} \frac{\exp(-t/\tau_{decay}^{NMDA}) - \exp(-t/\tau_{rise}^{NMDA})}{1 + 0.33\exp(-0.075V(t) - 2.1)}. \quad (8)$$

Here, g_{NMDA}, τ_{decay}^{NMDA}, and τ_{rise}^{NMDA} denote the parameters that determine the magnitude and kinetics of the NMDAR conductance (Eq. 4). The coefficient k_{sp} is proportional to the ratio of dendritic to spine membrane area and also depends on the Ca^{2+} permeability of NMDARs. We used k_{sp} = 22.2 so that the NMDAR-mediated Ca^{2+} response during postsynaptic depolarization (~16 μM at V = 0 mV) is around 10 times as large as the amplitude of the spike-evoked transient (~1.6 μM), similar to the data of Sabatini et al. (2002). The voltage dependence of the conductance in Eq. 8 is modified from Jahr and Stevens (1990).

2.4. Ca^{2+}-Dependent Plasticity Model

We applied the Ca^{2+}-dependent plasticity (CaDP) model, proposed in the previous study (Kubota and Kitajima 2008), to the simulations using the pairing protocol (e.g., Feldman (2000)) (Figs. 2 and 3). This model is based on the experimental findings that LTP is induced by a higher level of Ca^{2+} increase, while LTD requires a moderate level of prolonged Ca^{2+} transients (Yang et al. 1999; Mizuno et al. 2001; Taniike et al. 2008; Sjöström and Nelson 2002). The change in the synaptic weight Δw was determined as a function of the peak amplitude of the spine Ca^{2+} transient $[Ca]_{peak}$ and the duration of the Ca^{2+} elevation T_{Ca} (Kubota and Kitajima 2008):

$$\Delta w = f_P([Ca]_{peak}) + f_D([Ca]_{peak}) \cdot f_B(T_{Ca} - \hat{T}([Ca]_{peak})). \quad (9)$$

Here, T_{Ca} is the duration of the period during which the Ca^{2+} concentration in the spine is above a threshold σ_D, i.e., T_{Ca} = $\int_{-\infty}^{\infty} \Theta([Ca]_{sp}(t) - \sigma_D)dt$ with the Heaviside step function Θ ($\Theta(x)$ = 1 for $x > 0$ and $\Theta(x)$ = 0 for $x \leq 0$). Functions f_P (>0) and f_D (<0) are responsible for LTP and LTD induction at high and moderate Ca^{2+} levels, respectively (Fig. 2A). Function f_B is also the Heaviside step function (i.e., $f_B(x) = \Theta(x)$) and plays a role in blocking LTD induction when the duration T_{Ca} is shorter than a threshold value of $\hat{T}$. The threshold $\hat{T}$ is assumed to be an increasing function of $[Ca]_{peak}$, which may be associated with a cAMP-mediated signaling pathway that makes LTD induction difficult at higher Ca^{2+} levels (Kubota and Kitajima 2008; Lisman 1989).

The functions f_P and f_D are expressed as follows (Kubota and Kitajima 2008) (Fig. 2A):

$$f_P(x) = \begin{cases} \eta_P \left[-\left(\frac{x - \sigma_M}{\sigma_M - \sigma_P} \right)^2 + 1 \right]^2, & \text{for } x \in (\sigma_P, \sigma_M), \\ 0, & \text{otherwise}, \end{cases} \tag{10}$$

$$f_D(x) = \begin{cases} -\eta_D \left[-\left(\frac{2x - (\sigma_P + \sigma_D)}{\sigma_P - \sigma_D} \right)^2 + 1 \right]^2, & \text{for } x \in (\sigma_D, \sigma_P), \\ 0, & \text{otherwise}, \end{cases} \tag{11}$$

where the maximum levels of LTP and LTD are $\eta_P = 1.3$ and $\eta_D = 1$, respectively. The threshold Ca^{2+} levels required for the induction of LTP and LTD are $\sigma_P = 5$ μM and $\sigma_D = 2.5$ μM, respectively, and $\sigma_M = 9$ μM. The function $\hat{T}([Ca]_{peak})$ is simply assumed to be linear: $\hat{T} = k_{Ca}[Ca]_{peak}$ with k_{Ca} = 9.3 ms/μM.

2.5. Synaptic Modification by using Random Inputs

In the simulation examining the synaptic modification by STDP in the presence of random inputs (Figs. 4–9), the neuron receives inputs, generated by Poisson processes, from 4000 excitatory and 800 inhibitory synapses (Kubota and Kitajima 2008). Excitatory synapses were activated by either uncorrelated spike trains or two groups of correlated spike trains, while inhibitory synapses are activated by uncorrelated spike trains. All the uncorrelated inputs were generated by independent Poisson spike trains having firing rate of 3Hz. The correlation among inputs was introduced by the method similar to Song and Abbott (2001) as follows. The excitatory synapses were assumed to be comprised of two equally sized groups.

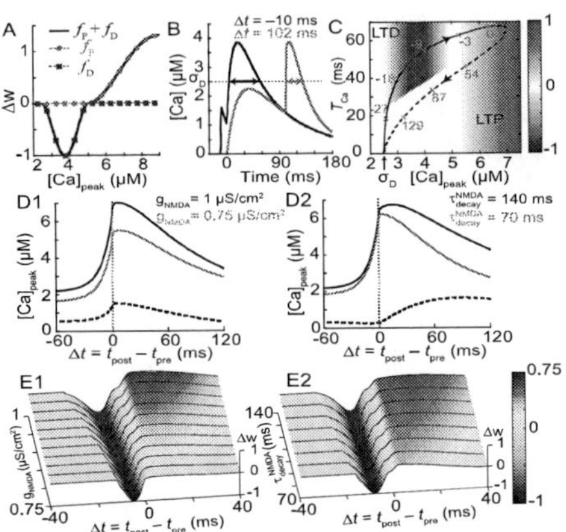

Figure 2. Predicted effects of changing NMDAR current properties on STDP. (A) Functions $f_P([Ca]_{peak})$ and $f_D([Ca]_{peak})$, for the CaDP model (Eq. 9), and their summation. (B) Calcium time courses when a presynaptic input is paired with postsynaptic firing for $\Delta t = -10\,\text{ms}$ (black) and 102 ms (red). Time is measured from the onset of presynaptic inputs. The bars with arrows denote the duration T_{Ca}, i.e., the time interval for which Ca2+ is above the threshold σ_D (2.5 μM). T_{Ca} is considerably longer for the post-pre timing ($T_{Ca} = 53\,\text{ms}$, black) than for the pre-post timing ($T_{Ca} = 28\,\text{ms}$, red) when the same Ca2+ peak level is attained (Kubota and Kitajima 2008). (C) T_{Ca} is plotted vs. $[Ca]_{peak}$ for each Δt. The arrows indicate the direction in which Δt increases; each number denotes its value. The background color shows the changes in plasticity Δw given by the CaDP model (Eq. 9), wherein LTD requires a relatively prolonged Ca2+ elevation (i.e., larger T_{Ca}) (Yang et al. 1999; Mizuno et al. 2001; Taniike et al. 2008; Sjöström and Nelson 2002). T_{Ca} is longer for $\Delta t < 0$ (solid line) than for $\Delta t > 0$ (dashed line) with the same Ca2+ peak level as in (B). Therefore, it is possible that the point $([Ca]_{peak}, T_{Ca})$ passes the LTD region for $\Delta t < 0$, but not for $\Delta t > 0$, inducing LTD only in the post-pre timing (Kubota and Kitajima 2008). The value of T_{Ca} decreases to 0 when $[Ca]_{peak} < \sigma_D$, from its definition. (D) The solid lines show $[Ca]_{peak}$ as function of Δt for two different values of g_{NMDA} (D1) and τ^{NMDA}_{decay} (D2). The dashed line shows the difference between the two solid lines. (E) The changes in the STDP curve with changes in g_{NMDA} (E1) or τ^{NMDA}_{decay} (E2). (Simulation parameters: $\tau^{NMDA}_{decay} = 90\,\text{ms}$ and $g_{NMDA} = 1\ \mu\text{S/cm}^2$ in (B) and (C), $\tau^{NMDA}_{decay} = 90\,\text{ms}$ in (D1) and (E1), and $g_{NMDA} = 0.9\ \mu\text{S/cm}^2$ in (D2) and (E2)).

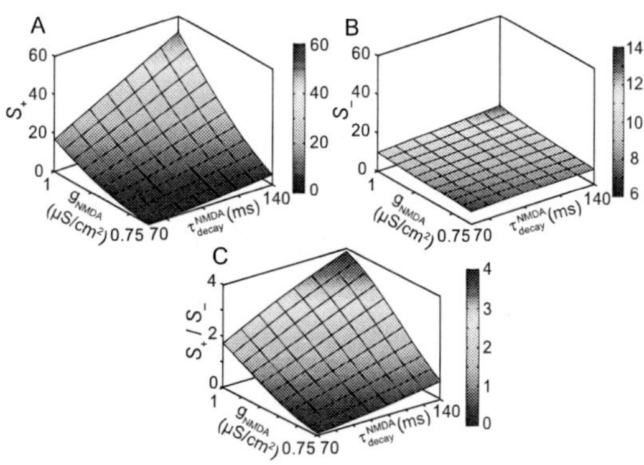

Figure 3. Changes in the areas under the STDP curve for the LTP (A) and LTD (B) portions (S_+ and S_-, respectively) and their ratio (S_+/S_-) (C) as function of g_{NMDA} and τ_{decay}^{NMDA}. Since the increase in S_+ caused by larger g_{NMDA} and τ_{decay}^{NMDA} is much more significant than that in S_-, the S_+/S_- ratio also increases with these parameters.

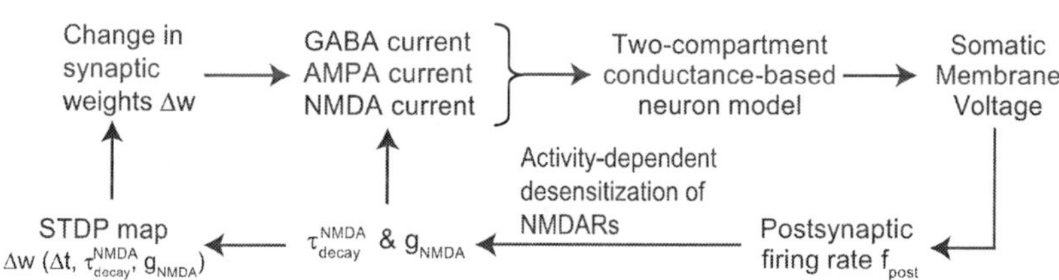

Figure 4. An STDP model incorporating metaplastic regulation by the ADFB mechanism. The conductance-based neuron receives random excitatory (AMPA and NMDA) and inhibitory (GABA) inputs. From the somatic membrane voltage of the neuron, the instantaneous firing rate (f_{post}) is estimated (Eq. 14), and the decay time constant and peak conductance of NMDARs (τ_{decay}^{NMDA} and g_{NMDA}, respectively) are dynamically modified through the activity-dependent desensitization (Eqs. 12 and 13). The changes in τ_{decay}^{NMDA} and g_{NMDA} are reflected in the NMDAR currents as well as the amount of plasticity Δw by using the STDP map generated from the simulation of the pairing protocol (Fig. 2). The weight changes caused by different spike pairs are linearly summed and the peak AMPAR conductances are modified accordingly.

We divided time into intervals taken from an exponential distribution with an average interval τ_c. For every time interval, we generated a random number y from a Gaussian distribution having mean 0 and standard deviation 1. The firing rate of the Poisson spikes for input neurons was set to $r_c(1+0.3y)$ (in Hz) and was fixed at this value until the onset of the next time interval. We used independent interval start times and independent values of y for the two groups so that they were uncorrelated with each other. The mean input frequency was set to r_c = 3 Hz, unless otherwise specified, and the correlation time was τ_c = 5 ms. Given a low success rate of synaptic transmission in cortical synapses (~10%) (Hessler et al. 1993), the input rate of 3 Hz will correspond to a presynaptic firing rate of ~30 Hz, which will be in the physiological range for the sensory-evoked response of cortical neurons (Ahmed et al. 1998).

STDP acts on all the excitatory synapses. To determine the amount of weight change Δw induced by each pre- and postsynaptic spike pair, we used a four-dimensional STDP map (see below) obtained from the simulation using the CaDP model (Fig. 4). The weight updating rule is assumed to be additive, and the weight changes caused by all the spike pairs are linearly summed. To stabilize the learning processes, both upper and lower bounds were imposed such that $w \in [0, w_{max}]$ with w_{max} = 2. The modification in synaptic weights was reflected in the peak AMPAR conductances as shown in Eq. 3.

2.6. Activity - and Subunit-Dependent Modulation of NMDARs

We considered the effects of the changes in NMDAR subunit expression on the synaptic parameters. In the forebrain, the NR2B-containing NMDARs are predominant at birth, whereas the number of NR2A-containing NMDARs increases over postnatal development, as mentioned above (Flint et al. 1997; Mierau et al. 2004; Monyer et al. 1994; Quinlan et al. 1999a, 1999b). In addition, the switching in the NR2 subunits takes place triggered by the changes in the firing activity or the neurotrophin level (Caldeira et al. 2007; Quinlan et al. 1999a, 1999b), suggesting that the expression pattern of distinct NR2 subunits may be regulated depending on different conditions.

Since this subunit switch has various effects on the physiological properties of NMDARs, we focused on the effects that are relatively well documented. First, the increase in the expression of NR2A-containing NMDARs significantly decreases the decay time constant of NMDAR-mediated synaptic currents (Flint et al. 1997; Mierau et al. 2004; Monyer et al. 1994). Next, intracellular Ca^{2+} induces the desensitization of NR2A-containing NMDARs but not that of NR2B-containing NMDARs (Krupp et al. 1996; Umemiya et al. 2001), suggesting that the subunit switch from NR2B to NR2A will strengthen the Ca^{2+}-dependent desensitization of NMDARs. Therefore, in the case that the postsynaptic activity increases, the accumulation of intracellular Ca^{2+}, via the frequent activation of voltage-gated Ca^{2+} channels (VGCCs) (Helmchen et al. 1996; Svoboda et al. 1997), will cause the desensitization of NMDARs with the strength dependent on the NR2 subunit composition. Additionally, considering the fact that the Ca^{2+}-dependent desensitization can decrease not only the peak conductance but also the decay time constant of NMDARs (Legendre et al. 1993; Medina et al. 1999; Rosenmund et al. 1995; Umemiya et al. 2001), we assumed that both of them are negatively regulated by the postsynaptic firing rate f_{post} as follows:

$$\tau_{decay}^{NMDA} = (1-\rho)\tau_0 + \rho\tau_1 - k_1\rho f_{post}, \quad (12)$$

$$g_{NMDA} = g_{NMDA}^0 - k_2\rho f_{post}. \quad (13)$$

Here, ρ $(0 \le \rho \le 1)$ is a non-dimensional parameter representing the state associated with the NMDAR subunit expression: ρ = 0 corresponds to a state where the NR2B-containing receptors are dominant as in neonatal neurons, whereas ρ = 1 denotes the state where the NMDARs comprise many NR2A subunits as in mature neurons. Thus, the first two terms on the right-hand side of Eq. 12 represents the acceleration of the decay kinetics of single NMDAR currents (i.e., at f_{post} = 0) by the switching from the NR2B to NR2A subunit (Flint et al. 1997; Mierau et al. 2004; Monyer et al. 1994). The decay time constants for ρ = 0 and 1 are set to τ_0 = 140 ms and τ_1 = 90 ms, respectively, from the data regarding the developmental change in the decay kinetics of NMDARs (Mierau et al. 2004). The peak conductance of single NMDAR currents was fixed at g_{NMDA}^0 = 1 μS/cm^2 independent of ρ, since there is evidence that the peak amplitude of single synaptic NMDAR currents does not change by the subunit switch (Carmignoto and Vicini 1992; Prybylowski et al. 2000; Shi et al. 1997). The last terms on the right-hand sides of these two equations describe the facilitation of the activity-dependent desensitization by the increase in the NR2A-containing NMDARs, as mentioned above. At ρ = 0, the postsynaptic activity does not influence τ_{decay}^{NMDA} and g_{NMDA}, whereas at ρ = 1, it decreases both these values with gradients of k_1 = 0.4 ms/Hz for τ_{decay}^{NMDA} and k_2 = 0.005 μS/(cm^2 Hz) for g_{NMDA}. These parameter values were selected such that the firing rate obtained by the simulation (Fig. 5C) (20–200 Hz) largely includes a range of cortical sensory-evoked responses (Ahmed et al. 1998).

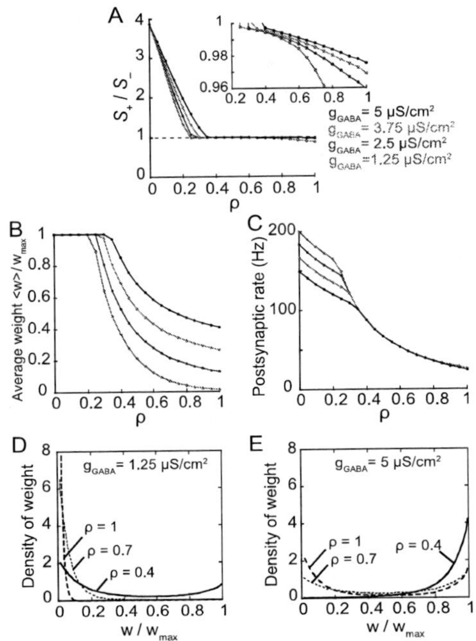

Figure 5. Equilibrium properties of STDP by using uncorrelated inputs. (A–C) The S_+/S_- ratio (temporally averaged value) (A), the average weight over all synapses (B), and the postsynaptic firing rate (C) at the equilibrium are shown for various values of ρ and g_{GABA} (g_{GABA} = 1.25 (green), 2.5 (blue), 3.75 (red), or 5 μS/cm2 (black)). Inset in (A): higher magnification of the S_+/S_- ratio. (D and E) The weight distributions at the steady state with ρ = 0.4 (solid), 0.7 (dotted), or 1 (dashed) for g_{GABA} = 1.25 (D) or 5 μS/cm2 (E).

The postsynaptic firing rate f_{post} was estimated by the following equation (Tanabe and Pakdaman 2001):

$$f_{post}(t) = \int_0^{\infty} \lambda \exp(-\lambda\tau) S_{post}(t-\tau) d\tau \,, \tag{14}$$

where $S_{post}(t) = \sum_f \delta(t - t_{post}^f)$ denotes the postsynaptic spike train and λ = 0.1 /s (λ^{-1} = 10 s) (Kubota and Kitajima 2010). Since the rise time (Carmignoto and Vicini 1992), the voltage dependence (Feldman et al. 1998; Kuner and Schoepfer 1996; Monyer et al. 1994; but also see Kato and Yoshimura 1993), and the Ca^{2+} permeability (Schneggenburger 1996) of NMDA currents are not significantly affected by the subunit change, the parameters linked to these receptor properties were maintained at constant values.

2.7. STDP Map

An STDP map was generated to examine the influence of NMDAR-dependent modulation of STDP on the synaptic dynamics (see Results). This map is four-dimensional and represented by Δw (Δt , τ_{decay}^{NMDA} , g_{NMDA}), where $\Delta t = t_{post} - t_{pre}$ denotes the interspike interval (ISI) between the pre- and postsynaptic events. To construct this map, we obtained the values of the synaptic weight change Δw corresponding to various sets of Δt , τ_{decay}^{NMDA} , and g_{NMDA} , by the simulation for the pairing protocol using the CaDP model (Fig. 2), and then multiplied them by a factor of 0.008. In this STDP map, Δt changes from –45 ms to 105 ms by 1 ms, τ_{decay}^{NMDA} changes from 70 ms to 140 ms by 2 ms, and g_{NMDA} changes from 0.75 μS/cm^2 to 1 μS/cm^2 by 0.005 μS/cm^2 (the total number of data points is about 280,000). The linear interpolation of the map data was employed to calculate Δw corresponding to each pre- and postsynaptic spike pair, from the values of Δt , τ_{decay}^{NMDA} , and

g_{NMDA}, where τ_{decay}^{NMDA} and g_{NMDA} were dynamically modified by Eqs. 12 and 13.

3. RESULTS

3.1. NMDAR-Mediated Modulation of STDP

Ca^{2+} signals in the spine have been suggested to be critical for controlling synaptic plasticity (Lisman 1989; Artola and Singer 1993). To investigate STDP on the basis of this Ca^{2+}-dependent plasticity (CaDP) hypothesis, we performed numerical experiments using the conductance-based pyramidal neuron (Fig. 1A), which includes the Ca^{2+} dynamics in the dendritic spine (see Methods).

A conventional model of CaDP is that higher and moderate levels of Ca^{2+} increase induce LTP and LTD, respectively (Lisman 1989; Artola and Singer 1993; Kitajima and Hara 2000; Karmarkar et al. 2002). However, this Ca^{2+} level-based model cannot reproduce the asymmetric time windows of STDP, characterized by LTP in the pre-post timing and LTD in the post-pre timing, observed in neocortical cells (Feldman 2000; Froemke and Dan 2002; Froemke et al. 2005). This is because according to this model, when a postsynaptic spike occurs a sufficiently long time after a presynaptic spike, Ca^{2+} levels drop low enough so that LTD occurs also in the pre-post timing (Rubin et al. 2005; Bi and Rubin 2005; Kubota and Kitajima 2008). Furthermore, accumulating evidence suggests that not only the Ca^{2+} level but also the duration of Ca^{2+} elevation is involved in the LTD induction: high Ca^{2+} elevation induces LTP, while a prolonged period of moderate Ca^{2+} increase is required for LTD (Yang et al. 1999; Mizuno et al. 2001; Taniike et al. 2008; Sjöström and Nelson 2002). Therefore, to incorporate these observations, the author used the previously proposed CaDP model (Kubota and Kitajima 2008), wherein the changes in the synaptic efficacy depend on both the peak amplitude and the duration of Ca^{2+} transients (see Methods).

To simulate the *in vitro* experiment of STDP (e.g., Feldman 2000), a presynaptic input was paired with a brief somatic current injection. The presynaptic input is composed of NMDAR- and AMPAR-mediated excitatory currents. GABA-mediated inhibition was not considered here because there is evidence that neocortical STDP is not affected by GABA activity (Froemke and Dan 2002). Ca^{2+} influx into the spine is mediated by both NMDARs and

VGCCs (Eq. 6). After calculating the Ca^{2+} time course for each ISI (Δt) between pre- and postsynaptic events, we applied the CaDP model given by Eq. 9. When the same Ca^{2+} peak level is attained, the duration of Ca^{2+} increase (T_{Ca}) is significantly shorter for the pre-post timing ($\Delta t > 0$) than for the post-pre timing ($\Delta t < 0$) (Fig. 2B), due to the characteristic of spike-induced Ca^{2+} influx via the Mg^{2+} block of NMDARs (Kubota and Kitajima 2008). Therefore, as depicted in Fig. 2C, the LTD block that depends on the duration of Ca^{2+} increase (f_B in Eq. 9) can prevent LTD for the pre-post timing while maintaining that for the post-pre timing (Kubota and Kitajima 2008), thereby reproducing STDP curves with asymmetric time windows (Fig. 2E).

To examine how the alterations in NMDAR current properties modulate STDP, we performed the simulations with changing the NMDAR peak conductance g_{NMDA} and its decay time constant τ_{decay}^{NMDA}. As expected, the increase in either g_{NMDA} or τ_{decay}^{NMDA} contributes to the accumulation of Ca^{2+} and increases $[Ca]_{peak}$ at all ISIs (Fig. 2D). However, the change in $[Ca]_{peak}$ is significantly larger for the pre-post timing than for the post-pre timing (Fig. 2D, dashed lines). Therefore, while LTP in the pre-post timing is considerably enhanced, LTD in the post-pre timing is less significantly affected (Fig. 2E). This was confirmed using the areas under the STDP curve for the LTP and LTD portions—$S_+ = \int_{f>0} f(x)\,dx$ and $S_- = -\int_{f<0} f(x)\,dx$—and their ratio S_+/S_-. As shown in Fig. 3, larger g_{NMDA} and τ_{decay}^{NMDA} increase S_+ much more than S_- (Figs. 3A and 3B) and therefore the S_+/S_- ratio increases (Fig. 3C), suggesting that stronger and more prolonged NMDAR activation produces a bias toward LTP. This result is an extension of our previous study (Kubota and Kitajima 2008) that shows that the prolongation of NMDAR activity augments S_+/S_-.

The higher sensitivity of LTP to the changes in NMDAR current properties results from a simple mechanism that the time course of NMDAR activation primarily affects the spike-induced Ca^{2+} influx during its activation, i.e., in the pre-post timing (Kubota and Kitajima 2008). Additionally, since the threshold Ca^{2+} level is higher for LTP than for LTD, moderately decreased Ca^{2+} responses inevitably eliminate the LTP region while maintaining a comparatively large LTD region (Fig. 2E). This means that S_+ is more susceptible than S_{-} to changes in the amplitude of Ca^{2+} transients around the threshold levels. Therefore, similar results will be obtained independent of the detail of the CaDP model: in fact, other CaDP models also suggest that the time course of NMDAR activation mainly affects LTP in the pre-post timing (Karmarkar and Buonomano 2002; Yeung et al. 2004).

3.2. Synaptic Distribution Regulated by NMDARs and GABA

The above findings suggest that the NMDARs can be involved in modifying the LTP/LTD balance in STDP. To investigate how this effect influences the dynamics of synaptic population, we simulated a case where the conductance-based neuron receives many random inputs from both excitatory (AMPA and NMDA) and inhibitory (GABA) synapses (see Methods). The numbers of excitatory and inhibitory synapses (4000 and 800, respectively) were selected such that they approximately agree with the data of developing neocortical neurons (Micheva and Beaulieu 1996). All the synapses were activated by uncorrelated spike trains with the same frequency (3 Hz) (see Methods).

When multiple random inputs drive a neuron, the postsynaptic Ca^{2+} would have complicated waveforms consisting of a number of local peaks, which makes it quite difficult to directly apply the Ca^{2+} peak level- and duration-dependent plasticity model (Eq. 9) (Kubota and Kitajima 2008). Therefore, to examine the possible effects of dynamically modulating STDP through NMDARs, we constructed an STDP map, represented as $\Delta w(\Delta t, \tau_{decay}^{NMDA}, g_{NMDA})$, by using the results of the previous simulation employing the pairing protocol (Fig. 2) (see Methods). The linear interpolation of the map data was used to calculate the weight change Δw induced by each pre- and postsynaptic spike pair, from its ISI Δt, τ_{decay}^{NMDA}, and g_{NMDA} (Fig. 4). The values of τ_{decay}^{NMDA} and g_{NMDA} were temporally modified by the ADFB mechanism given by Eqs. 12 and 13, which models the activity- and subunit-dependent desensitization of NMDARs (Flint et al. 1997; Krupp et al. 1996; Legendre et al. 1993; Medina et al. 1999; Mierau et al. 2004; Monyer et al. 1994; Rosenmund et al. 1995; Umemiya et al. 2001) with the parameter ρ representing the state of the NMDAR subunit expression.

Figure 5 shows the equilibrium properties of STDP (i.e., the state where the synaptic weights converge to a stationary distribution) for various values of ρ and the peak GABA conductance g_{GABA}. In the presence of ADFB, the values of τ_{decay}^{NMDA} and g_{NMDA} decline with higher postsynaptic activity (Eqs. 12 and 13) and, furthermore, the decrease in these synaptic parameters significantly decreases the S_+ / S_- ratio, as expected from Fig. 3C. Therefore, the enhanced firing activity generated by stronger inputs can decrease the overall effect of potentiation as compared to that of depression, producing an approximate balance between synaptic strengthening and weakening (Kubota et al. 2009; Tegnér and Kepecs 2002). Therefore, the temporal average of S_+ / S_- converges to a value slightly smaller than 1 for relatively larger ρ values (Fig. 5A), although the time course of S_+ / S_- is highly irregular (Fig. 6).

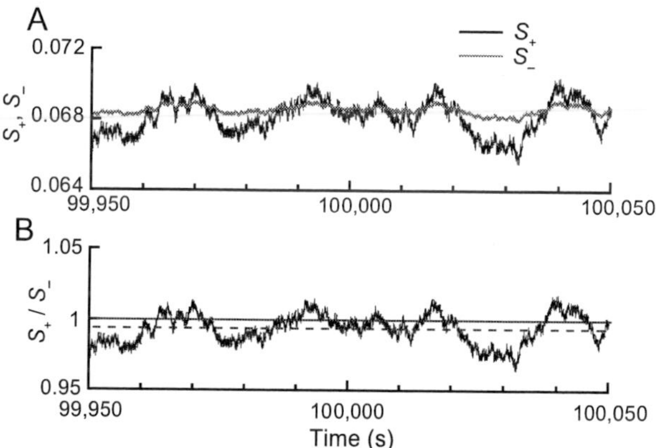

Figure 6. Time courses of the areas of the STDP curve under the LTP and LTD portions (S_+ and S_-, respectively) (A) and their ratio (S_+ / S_-) (B) at the equilibrium of STDP in the presence of uncorrelated inputs. The dashed line in (B) denotes the temporally averaged value of the S_+ / S_- ratio (0.994) (ρ = 0.5 and g_{GABA} = 5 µS/cm2).

Although further increase in ρ will act to decrease S_+ / S_-, a small decrease in S_+ / S_- significantly decreases the average weight and postsynaptic activity (Figs. 5B and 5C) (Song et al. 2000), through the reduction in the number of synapses gathering at the upper limit (Figs. 5D and 5E). Therefore, the facilitation of the ADFB modulation by the increase in ρ can be counterbalanced by the decrease in the postsynaptic activity, so that the S_+ / S_- ratio is maintained at nearly constant values for larger ρ (Fig. 5A, inset).

In contrast to the strong impact of ρ on the postsynaptic firing rate, changing g_{GABA} does not significantly affect it (Fig. 5C). Instead, enhancing GABA activity augments the average weight (Fig. 5B) by slightly increasing the S_+ / S_- ratio (Fig. 5A, inset) and pushing more synapses toward the upper limit (Figs. 5D and 5E). This indicates the existence of strong regulatory function intrinsic to ADFB that keeps excitatory-inhibitory balance through the precise control of LTP/LTD balance (Kubota et al. 2009).

3.3. Correlation-Based Competition Regulated by NMDARs and GABA

The convergence to a bimodal weight distribution (Figs. 5D and 5E) indicates the existence of competitive interaction among inputs (Song et al. 2000). We next explored how the competitive function of STDP can regulate the synaptic dynamics when a neuron receives two groups of correlated inputs, similar to visual cortical neurons receiving inputs from the two eyes. Excitatory synapses were divided into two equally sized groups (with 2000 inputs each) and independent input correlations were applied to each of them with the same magnitude (Song and Abbott 2001) (See Methods). When the level of GABA inhibition was relatively weak (g_{GABA} = 1.25 μS/cm^2; Fig. 7A), the two groups of synapses was found to converge to the identical average weight independent of ρ. In contrast, when the GABA activity became sufficiently large (g_{GABA} = 5 μS/cm^2; Fig. 7B), the synaptic weights segregated into the two input groups with the one winning the competition inhibiting the other for larger ρ values (which group wins is random). This result was clarified by the weight distribution at the stationary state (Fig. 8), which shows that the two synaptic groups have the same distributions when either ρ or g_{GABA} is small, and that the segregation of weight distributions requires both large ρ and large g_{GABA}.

To quantify the level of competitive interaction between the correlated input groups, we introduced two measures: the difference in average weights normalized by the maximum weight and the synaptic competition index (SCI) (Kubota and Kitajima 2010), which are described as $|<w_1> - <w_2>|/w_{max}$ and $|<w_1> - <w_2>|/(<w_1> + <w_2>)$, respectively, with the steady-state average weight $<w_i>$ for group i. An SCI of 0 indicates that the neuron is equally responsive to the two input groups, whereas an SCI of 1 means that the cell response is dominated by either of the two groups. The index analogous to SCI is frequently employed by physiological experiments of ocular dominance plasticity (e.g., Rittenhouse et al. 2006) to quantify the relative contribution of each eye to the visual cell response.

When either ρ or g_{GABA} is small, both measures of competition are 0 (Figs. 7C and 7D), showing that both groups have the same average strength at the steady state. However, when both ρ and g_{GABA} are sufficiently large (ρ>~0.6, g_{GABA}>~2.5μS/cm^2), the two measures take positive values, indicating that the two groups segregate into the ones winning the competition and losing it. The S_+/S_- ratio and postsynaptic rate are gradually altered by the changes in ρ and g_{GABA} similar to the uncorrelated input case (Figs. 7E and 7F), implying that the emergence of the between-group competition does not influence the LTP/LTD balance or total afferent activity. The results here suggest that the induction of competition requires both the strengthening of ADFB function, provided by the mature state of NMDAR subunit expression (larger ρ), as well as the enhancement of GABA inhibition (larger g_{GABA}).

Similar changes in the competitive properties with changes in ρ and g_{GABA} were shown to occur by the widely-used exponential-type STDP curve incorporated with ADFB (Kubota and Kitajima 2010). Further, similar results were found to occur when the firing dynamics of a neuron was reduced to the leaky integrate-and-fire model (data not shown). Therefore, our results are independent of the detail of the STDP map or the firing mechanism. We have also found the change in the input-output correlation function similar to the previous study (Kubota and Kitajima 2010) such that larger ρ tends to increase the temporal width of the correlation while the larger g_{GABA} tends to increase its peak amplitude (data not shown). This means that larger ρ and larger g_{GABA} act cooperatively to strengthen the input-output correlation, which will permit STDP to reflect the correlation structure among inputs (Kubota and Kitajima 2010).

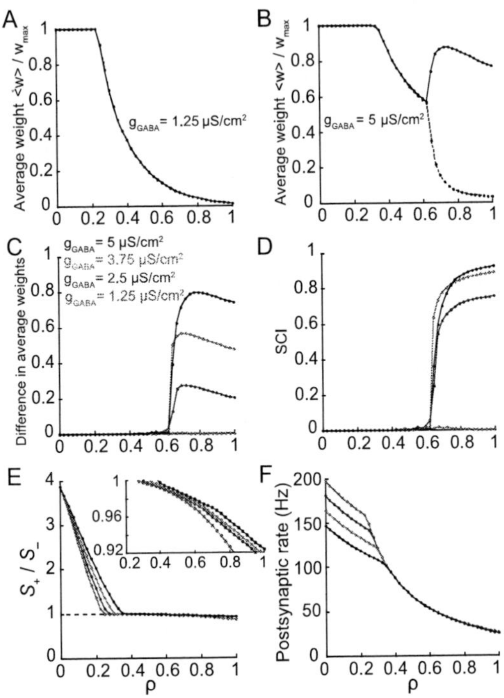

Figure 7. Equilibrium properties of STDP in the presence of two groups of correlated inputs. (A and B) Average synaptic weights of the two input groups at the equilibrium state as function of ρ, where $g_{GABA} = 1.25$ (A) or 5 μS/cm2 (B). Both the groups have the same average weight for all ρ values in (A), while they segregate into strong and weak ones at larger ρ values in (B). (C and D) The difference in average weights between the two groups (normalized by the maximum weight) (C) and the values of SCI (D) for various values of ρ and g_{GABA}. (E and F) The changes in the S_+ / S_- ratio (temporally averaged value) (E) and the postsynaptic firing rate (F) as function of ρ and g_{GABA}. Inset in (E): Higher magnification of the S_+ / S_- ratio. (g_{GABA} = 1.25 (green), 2.5 (blue), 3.75 (red), or 5 μS/cm2 (black) in (C)–(F).)

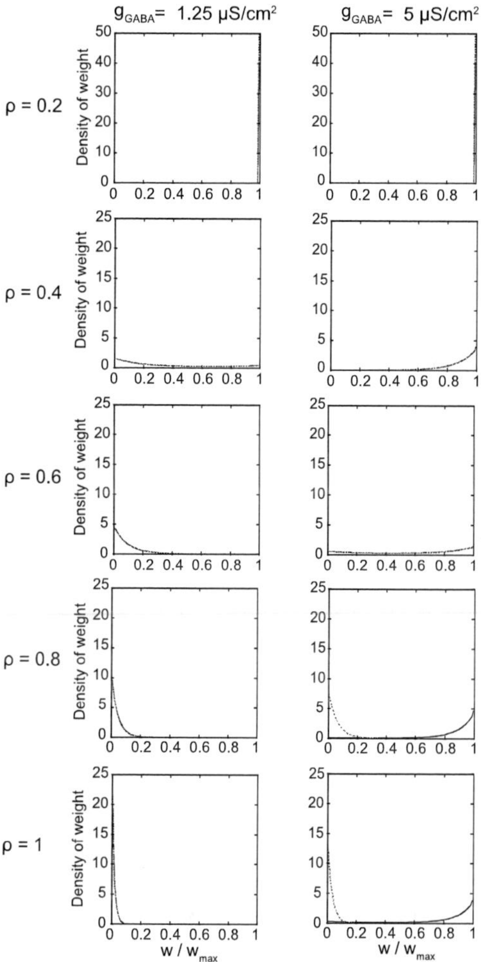

Figure 8. Steady-state weight distributions generated by STDP in the presence of two groups of correlated inputs. The weight distributions of the two groups are shown by black solid and red dashed lines for various values of ρ and g_{GABA}. In the cases that GABA inhibition is relatively weak (g_{GABA} = 1.25 μS/cm2, left column), the weight distributions of the two groups are identical for all ρ values. However, for stronger GABA inhibition (g_{GABA} = 5 μS/cm2, right column), the competition between the different groups occurs for sufficiently large ρ values (ρ = 0.8 and 1) and the synaptic efficacies segregate into the two input groups.

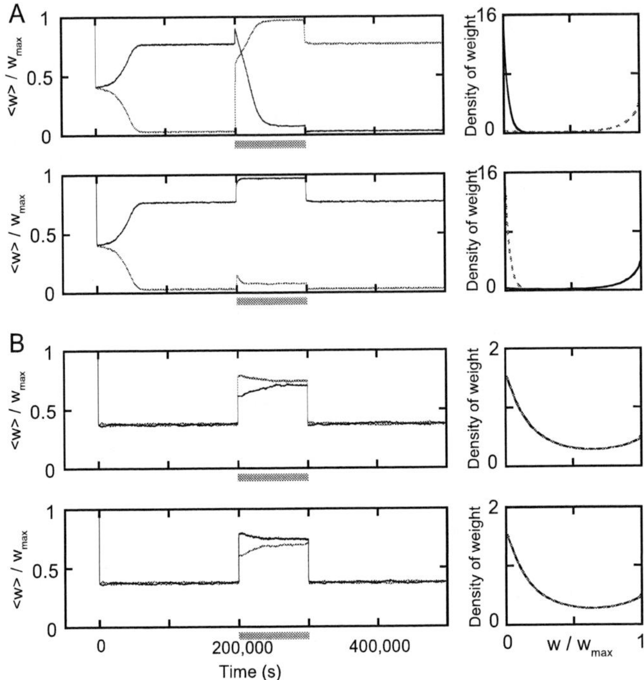

Figure 9. Experience-dependent synaptic modifications emerging through the competition between the input groups. Left column: the time courses of the average weights of the two groups are depicted by black and red lines. After the input frequency for all the synapses are maintained at 3 Hz, that for the group represented by black (A (top) and B (top)) or red (A (bottom) and B (bottom)) line is decreased to 1.5 Hz for 200,000 < t < 300,000 (in s) (gray bar) and then recovered to 3 Hz. Right column: the final weight distributions for both the groups by using the same line colors as those in the left column. (A) For sufficiently large values of ρ and g_{GABA} (ρ = 1, g_{GABA} = 5μS/cm2), where the competition arises between the different groups, the dominant group at the final state becomes the one whose activity has not been suppressed, regardless of which group receives the suppression of input activity. (B) For smaller values of ρ and g_{GABA} (ρ = 0.4, g_{GABA} = 1.25 μS/cm2), where the competition does not take place, the final weight distributions are identical for both the groups; therefore, the information regarding the history of sensory inputs cannot be retained. Accordingly, in the presence (A), but not in the absence (B), of competitive interaction, the learned synaptic weights can reflect the experience of earlier input activities.

To examine functional consequences of the competition between the correlated groups, I performed simulations resembling monocular deprivation by transiently suppressing the activation frequency of either one synaptic group. Figure 9A shows the time course of the mean weights (left column) and the final weight distribution (right column) of the two groups, when sufficiently large values of ρ and g_{GABA} induce between-group competition. After the mean activation frequency for the inputs of both groups was maintained at 3 Hz, the frequency for the group winning (Fig. 9A, top) or losing (Fig. 9A, bottom) the competition was decreased by 50% for 100,000 s (denoted by gray bar) and then again recovered to 3 Hz. The activation rate was changed stepwise, so that the decrease (increase) in the total afferent activity rapidly increases (decreases) the S_+ / S_- ratio through the ADFB modulation.

Therefore, the synaptic weights of the two groups are increased and decreased rapidly at the onset and end of the input frequency modification, respectively, in both the cases shown in Fig. 9A.

However, only when the input activity of the winning group is suppressed (Fig. 9A, top), the dominant group is switched into the one that has been losing before the input frequency is modified. Accordingly, the final synaptic distribution is determined by the past experience of inputs such that the synaptic group that has not been suppressed dominates over the one that has been suppressed, independent of which group receives the suppression of inputs (Fig. 9A, right column). This result is reminiscent of the experimental observations in ocular dominance plasticity, wherein the response of the visual cells is dominated by the non-deprived eye after the monocular deprivation (Wiesel 1982; Gordon and Stryker 1996). It is important to note that such input experience-dependent synaptic modification does not occur when the between-group competition is absent because of lower ρ and/or g_{GABA}, as shown in the example of Fig. 9B. In this case, the two groups converge to the same weight distribution independent of the history of input activities. This is again reminiscent of the neurons in the pre critical period where the neuronal responses cannot reflect the past sensory experience. The results similar to Fig. 9 were also obtained when the duration of the input rate modification was relatively shorter (10,000 s) or longer (500,000 s) than the time scale of the change in the synaptic pattern (data not shown). The findings here suggest that the correlation-based competition between the input groups, caused by larger

ρ and g_{GABA}, would be required for the synaptic pattern to reflect the information regarding the past sensory experience during the critical period.

4. Discussion

4.1. Role of NMDARs and GABA in Developmental Plasticity

Although NMDAR-dependent LTP and LTD are hypothesized to underlie sensory experience-dependent plasticity (Crair and Malenka 1995; Roberts and Ramoa 1999; Erisir and Harris 2003; Dumas 2005), this idea is inconsistent with the finding that enhancing cortical GABA triggers ocular dominance plasticity in the visual cortex (Hensch et al. 1998; Fagiolini and Hensch 2000; Hensch 2005). This effect of GABA *in vivo* appears to be difficult to understand because GABA activity does not influence STDP *in vitro* in the visual cortical slices (Froemke and Dan 2002). In this study, to explore an integrative role of NMDARs and GABA in developmental plasticity, I first investigated how STDP can be modulated depending on NMDARs, based on the Ca^{2+}-based hypothesis of plasticity (Kubota and Kitajima 2008) (Fig. 2). It has been shown that the increase in the NMDAR peak conductance as well as in its decay time constant primarily facilitates LTP more significantly than LTD, increasing the S_+ / S_- ratio (Fig. 3). The strong dependence of LTP on the NMDAR activation not only agrees with previous theoretical models (Karmarkar and Buonomano 2002; Yeung et al. 2004) but also is in line with the experimental observations that indicate the involvement of postsynaptic NMDARs particularly in LTP in STDP (Caporale and Dan 2008).

When the neuron receives random inputs, as in *in vivo* conditions, the ADFB function can produce a dynamic balance between LTP and LTD (Fig. 5A) (Tegnér and Kepecs 2002; Kubota et al. 2009). In the presence of this LTP/LTD balance, STDP acts to approximately keep the balance between excitation and inhibition (Song et al. 2000). Therefore, although GABA activity is not involved in STDP *in vitro*, the enhanced GABA inhibition could function to strengthen excitation *in vivo*, facilitating LTP of excitatory synapses (Fig. 5B). This means that inhibition may be able to indirectly control plasticity through the regulatory action of STDP. In the presence of two groups of correlated inputs, which resemble inputs from the two eyes to a visual cortical cell, STDP exhibits input correlation-based competition in the coexistence of two factors: the mature state of NMDAR subunit expression (larger ρ), which enhances ADFB via the NR2A-containing NMDARs exhibiting Ca^{2+}-dependent desensitization (Krupp et al. 1996; Umemiya et al. 2001), and the increased level of background GABA inhibition (larger g_{GABA}) (Figs. 7C and 7D). This result, obtained by taking into account the CaDP mechanism, further supports the hypothesis of a cooperative role of NMDARs and GABA in introducing correlation-based competition, which has been proposed in the previously study (Kubota and Kitajima, 2010). The competitive interaction can produce a "memory" of the past sensory inputs (Angeli et al. 2004; Shpiro et al. 2007), because which input group becomes dominant depends on earlier input activities (Fig. 9), and therefore will provide the basis of experience-dependent plasticity in the critical period. In fact, the involvement of activity-dependent competition in experience-dependent cortical plasticity has been suggested by many experimental observations (Rauschecker and Singer 1979; Wiesel 1982; Shatz 1990; Gordon and Stryker 1996).

4.2. Ca^{2+} Signaling in STDP Mediated by NMDARs

The mechanism for plasticity in the present model is based on the Ca^{2+} peak level- and duration-dependent model (Kubota and Kitajima 2008), where the LTP induction requires a higher spine Ca^{2+} level, whereas the LTD induction necessitates both an intermediate Ca^{2+} level and prolonged duration of Ca^{2+} elevation (Eq. 9). This model may be considered as an extension of the conventional Ca^{2+} level-based model (Artola and Singer 1993; Cho et al. 2001; Cormier et al. 2001; Karmarkar et al. 2002; Kitajima and Hara 2000; Lisman 1989) toward the inclusion of the involvement of temporal factors of Ca^{2+} transients in the LTD induction (Mizuno et al. 2001; Sjöström and Nelson 2002; Taniike et al. 2008; Yang et al. 1999). The dependence of STDP on the NMDAR activation and the resulting rise in the intracellular Ca^{2+} level has been supported by many experiments (Bi and Poo 1998; Caporale and Dan 2008; Debanne et al. 1998; Feldman 2000; Magee and Johnston 1997; Markram et al. 1997; Sjöström et al. 2001; Zhang et al. 1998). Further, the requirement of the peak Ca^{2+} level greater than the distinct threshold levels for LTP and LTD agrees with recent experimental findings in STDP (Nevian and Sakmann 2006).

It would be important to note that although the CaDP hypothesis similar to our model is supported by many experiments, the plasticity induction protocols used in these experiments are generally simple, i.e., they basically consist of the repetition of a small number of spikes and/or bursts. Therefore, it would be still inconclusive whether the use of the CaDP hypothesis alone is sufficient to predict plasticity induced by more complicated firing pattern similar to natural conditions. In fact, a recent theoretical study (Shah et al. 2006) suggests that the inclusion of rapid short-term feedback effects, such as short-term depression of EPSPs and the attenuation of backpropagating action potentials caused by preceding spikes, may be required to make the predicted outcomes of CaDP model (Shouval et al. 2002) closer to the data of the STDP experiment examining multispike interaction (Froemke and Dan 2002). In the future study, it would be of interest to study weight modification dynamics through STDP by integrating the proposed ADFB mechanism with the CaDP incorporating the rapid short-term feedback effects (Shah et al. 2006).

Recently, it has been reported that the interaction between Ca^{2+} signaling mediated through different NR2 subunits may function to control the direction of synaptic modification in STDP (Gerkin et al. 2007). This experiment indicates that the activation of NR2A- and NR2B-containing NMDARs facilitates the LTP and LTD induction, respectively. Because the Ca^{2+}-dependent desensitization occurs in NR2A-containing NMDARs but not in those containing NR2B (Krupp et al. 1996; Umemiya et al. 2001), it can be expected from this experimental finding that LTP rather than LTD would be mainly subject to modulation via the activity-dependent desensitization of NMDARs. Therefore, although the subunit-specific control model of plasticity (Gerkin et al. 2007) is quite different from the assumption in this study, a key result in the present study that the Ca^{2+}-dependent NMDAR desensitization decrease the S_+ / S_- ratio will still remain valid in this case. Thus, the notion that dynamic LTP/LTD balance is achieved by the NMDAR function would be still consistent with the hypothesis of NR2 subunit-specific control of LTP and LTD.

Acknowledgment

This study is partially supported by Grant-in-Aid for Scientific Research (KAKENHI (19700281), Young Scientists (B)) from the Japanese government.

References

Ahmed, B; Anderson, JC; Douglas, RJ; Martin, KAC; Whitteridge, D (1998). Estimates of the net excitatory currents evoked by visual stimulation of identified neurons in cat visual cortex. *Cerebral Cortex, 8,* 462–476.

Angeli, D; Ferrell, JE; Sontag, ED (2004). Detection of multistability, bifurcations, and hysteresis in a large class of biological positive-feedback systems. *Proceedings of the National Academy of Sciences of the United States of America, 101,* 1822–1827.

Artola, A; Singer, W (1993). Long-term depression of excitatory synaptic transmission and its relationship to long-term potentiation. *Trends in Neurosciences, 16*, 480–487.

Bernander, O; Douglas, RJ; Martin, KAC; Koch, C (1991). Synaptic background activity influences spatiotemporal integration in single pyramidal cells. *Proceedings of the National Academy of Sciences of the United States of America, 88*, 11569–11573.

Bi, GQ; Poo, MM (1998). Synaptic modifications in cultured hippocampal neurons: Dependence on spike timing, synaptic strength, and postsynaptic cell type. *Journal of Neuroscience, 18*, 10464–10472.

Bi, GQ; Rubin, J (2005). Timing in synaptic plasticity: From detection to integration. *Trends in Neurosciences* 28: 222–228.

Caldeira, MV; Melo, CV; Pereira, DB; Carvalho, RF; Carvalho, AL; Duarte, CB (2007). BDNF regulates the expression and traffic of NMDA receptors in cultured hippocampal neurons. *Molecular and Cellular Neuroscience, 35*, 208–219.

Caporale, N; Dan, Y (2008). Spike timing-dependent plasticity: A Hebbian learning rule. *Annual Review of Neuroscience, 31*, 25–46.

Carmignoto, G; Vicini, S (1992). Activity-dependent decrease in NMDA receptor responses during development of the visual cortex. *Science, 258*, 1007–1011.

Cho, K; Aggleton, JP; Brown, MW; Bashir, ZI (2001). An experimental test of the role of postsynaptic calcium levels in determining synaptic strength using perirhinal cortex of rat. *Journal of Physiology*, 532.2, 459–466.

Cormier, RJ; Greenwood, AC; Connor, JA (2001). Bidirectional synaptic plasticity correlated with the magnitude of dendritic calcium transients above a threshold, *Journal of Neurophysiology*, 85, 399–406.

Crair, MC; Malenka, RC (1995). A critical period for long-term potentiation at thalamocortical synapses. *Nature, 375*, 325–328.

Daw, MI; Scott, HL; Isaac, JTR (2007). Developmental synaptic plasticity at the thalamocortical input to barrel cortex: Mechanisms and roles. *Molecular and Cellular Neuroscience, 34*, 493–502.

Debanne D; Gähwiler, BH; Thompson, SM (1998). Long-term synaptic plasticity between pairs of individual CA3 pyramidal cells in rat hippocampal slice cultures. *Journal of Physiology (London), 507.1,* 237–247.

Dumas, TC (2005). Developmental regulation of cognitive abilities: Modified composition of a molecular switch turns on associative learning. *Progress in Neurobiology, 76*, 189–211.

Erisir, A; Harris, JL (2003). Decline of the critical period of visual plasticity is concurrent with the reduction of NR2B subunit of the synaptic NMDA receptor in layer 4. *Journal of Neuroscience,* 23, 5208–5218.

Fagiolini, M; Hensch, TK (2000). Inhibitory threshold for critical-period activation in primary visual cortex. *Nature, 404,* 183–186.

Feldman, DE (2000). Timing-based LTP and LTD at vertical inputs to layer II/III pyramidal cells in rat barrel cortex. *Neuron, 27*, 45–56.

Feldman, DE; Nicoll, RA; Malenka, RC; Isaac, JTR (1998). Long-term depression at thalamocortical synapses in developing rat somatosensory cortex. *Neuron, 21*, 347–357.

Flint, AC; Maisch, US; Weishaupt, JH; Kriegstein, AR; Monyer, H (1997). NR2A subunit expression shortens NMDA receptor synaptic currents in developing neocortex. *Journal of Neuroscience, 17*, 2469–2476.

Froemke, RC; Dan, Y (2002). Spike-timing-dependent synaptic modification induced by natural spike trains. *Nature, 416*, 433–438.

Froemke, RC; Poo, MM; Dan, Y (2005). Spike-timing-dependent synaptic plasticity depends on dendritic location. *Nature, 434*, 221–225.

Gerkin, RC; Lau, PM; Nauen, DW; Wang, YT; Bi, GQ (2007). Modular competition driven by NMDAR receptor subtypes in spike-timing-dependent plasticity. *Journal of Neurophysiology, 97,* 2851–2862.

Gordon, JA; Stryker, MP (1996). Experience-dependent plasticity of binocular responses in the primary visual cortex of the mouse. *Journal of Neuroscience, 16*, 3274–3286.

Hanover, JL; Huang, ZJ; Tonegawa, S; Stryker, MP (1999). Brain-derived neurotrophic factor overexpression induces precocious critical period in mouse visual cortex. *Journal of Neuroscience, 19*, RC40.

Harris, KM; Jensen, FE; Tsao, B (1992). Three-dimensional structure of dendritic spines and synapses in rat hippocampus (CA1) at postnatal day

15 and adult ages: Implications for the maturation of synaptic physiology and long-term potentiation. *Journal of Neuroscience, 12*, 2685–2705.

Helmchen, F; Imoto, K; Sakmann, B (1996). Ca^{2+} buffering and action potential-evoked Ca^{2+} signaling in dendrites of pyramidal neurons. *Biophysical Journal, 70*, 1069–1081.

Hensch, TK (2005). Critical period plasticity in local cortical circuits. *Nature Reviews Neuroscience, 6,* 877–888.

Hensch, TK; Fagiolini, M; Mataga, N; Stryker, MP; Baekkeskov, S; Kash, SF (1998). Local GABA circuit control of experience-dependent plasticity in developing visual cortex. *Science, 282*, 1504–1508.

Hessler, NA; Shirke, AM; Mallnow, R (1993). The probability of transmitter release at a mammalian central synapse. *Nature, 366,* 569–572.

Huang, ZJ; Kirkwood, A; Pizzorusso, T; Porciatti, V; Morales, B; Bear, MF; Maffei, L; Tonegawa, S (1999). BDNF regulates the maturation of inhibition and the critical period of plasticity in mouse visual cortex. *Cell, 98*, 739–755.

Jahr, CE; Stevens, CF (1990). Voltage dependence of NMDA-activated macroscopic conductances predicted by single-channel kinetics. *Journal of Neuroscience, 10*, 3178–3182.

Karmarkar, UR; Buonomano, DV (2002). A model of spike-timing dependent plasticity: One or two coincidence detectors? *Journal of Neurophysiology, 88*, 507–513.

Karmarkar, UR; Najarian, MT; Buonomano, DV (2002) Mechanisms and significance of spike-timing dependent plasticity. *Biological Cybernetics,* 87, 373–382.

Kato, N; Yoshimura, H (1993). Reduced Mg^{2+} block of N-methyl-D-aspartate receptor-mediated synaptic potentials in developing visual cortex. *Proceedings of the National Academy of Sciences of the United States of America, 90*, 7114–7118.

Kirkwood, A; Bear, MF (1994). Hebbian synapses in visual cortex. *Journal of Neuroscience, 14*, 1634–1645.

Kitajima, T; Hara, K (2000). A generalized Hebbian rule for activity-dependent synaptic modifications. *Neural Networks, 13*, 445–454.

Koch, C. (1999). *Biophysics of computation.* New York: Oxford University Press.

Koch, C; Poggio, T (1983). A theoretical analysis of electrical properties of spines. *Proceedings of Royal Society of London B, 218*, 455–477.

Krupp, JJ; Vissel, B; Heinemann, SF; Westbrook, GL (1996). Calcium-dependent inactivation of recombinant N-methyl-D-aspartate receptors is NR2 subunit specific. *Molecular Pharmacology 50*, 1680–1688.

Kubota, S; Kitajima, T (2008). A model for synaptic development regulated by NMDA receptor subunit expression. *Journal of Computational Neuroscience, 24*, 1–20.

Kubota, S; Kitajima, T (2010). Possible role of cooperative action of NMDA receptor and GABA function in developmental plasticity, *Journal of Computational Neuroscience,* 28, 347–359.

Kubota, S., Rubin, J; Kitajima, T (2009). Modulation of LTP/LTD balance in STDP by an activity-dependent feedback mechanism, *Neural Networks,* 22, 527–535.

Kuner, T; Schoepfer, R (1996). Multiple structural elements determine subunit specificity of Mg^{2+} block in NMDA receptor channels. *Journal of Neuroscience, 16*, 3549–3558.

Legendre, P; Rosenmund, C; Westbrook, GL (1993). Inactivation of NMDA channels in cultured hippocampal neurons by intracellular calcium. *Journal of Neuroscience, 13*, 674–684.

Lisman, J (1989). A mechanism for the Hebb and the anti-Hebb processes underlying learning and memory. *Proceedings of the National Academy of Sciences of the United States of America, 86,* 9574–9578.

Magee, JC; Johnston, D (1997). A synaptically controlled, associative signal for Hebbian plasticity in hippocampal neurons. *Science, 275*, 209–213.

Markram, H; Lubke, J; Frotscher, M; Sakmann, B (1997). Regulation of synaptic efficacy by coincidence of postsynaptic APs and EPSPs. *Science, 275*, 213–215.

McCormick, DA; Connors, BW; Lighthall, JW; Prince, DA (1985). Comparative electrophysiology of pyramidal and sparsely spiny stellate neurons of the neocortex. *Journal of Neurophysiology, 54*, 782–806.

Medina, I; Leinekugel, X; Ben-Ari, Y (1999). Calcium-dependent inactivation of the monosynaptic NMDA EPSCs in rat hippocampal neurons in culture. *European Journal of Neuroscience, 11*, 2422–2430.

Micheva, KD; Beaulieu, C (1996). Quantitative aspects of synaptogenesis in the rat barrel field cortex with special reference to GABA circuitry. *Journal of Comparative Neurology, 373,* 340–354.

Mierau, SB; Meredith, RM; Upton, AL; Paulsen, O (2004). Dissociation of experience-dependent and -independent changes in excitatory synaptic transmission during development of barrel cortex. *Proceedings of the*

National Academy of Sciences of the United States of America, 101, 15518–15523.

Mizuno, T; Kanazawa, I; Sakurai, M (2001). Differential induction of LTP and LTD is not determined solely by instantaneous calcium concentration: An essential involvement of a temporal factor. *European Journal of Neuroscience, 14,* 701–708.

Monyer, H; Burnashev, N; Laurie, DJ; Sakmann, B; Seeburg, PH (1994). Developmental and regional expression in the rat brain and functional properties of four NMDA receptors. *Neuron, 12*, 529–540.

Nevian, T; Sakmann, B (2006). Spine Ca^{2+} signaling in spike-timing-dependent plasticity. *Journal of Neuroscience, 26*, 11001–11013.

Prybylowski, K; Fu, Z; Losi, G; Hawkins, LM; Luo, J; Chang, K; Wenthold, RJ; Vicini, S (2002). Relationship between availability of NMDA receptor subunits and their expression at the synapse. *Journal of Neuroscience, 22*, 8902–8910.

Quinlan, EM; Olstein, DH; Bear, MF (1999a) Bidirectional, experience-dependent regulation of N-methyl-D-aspartate receptor subunit composition in the rat visual cortex during postnatal development. *Proceedings of the National Academy of Sciences of the United States of America, 96*, 12876–12880.

Quinlan, EM; Philpot, BD; Huganir, RL; Bear, MF (1999b). Rapid, experience-dependent expression of synaptic NMDA receptors in visual cortex in vivo. *Nature Neuroscience, 2*, 352–357.

Rauschecker, JP; Singer, W (1979). Changes in the circuitry of the kitten visual cortex are gated by postsynaptic activity. *Nature, 280*, 58–60.

Rittenhouse, CD; Siegler, BA; Voelker, CA; Shouval, HZ; Paradiso, MA; Bear, MF (2006). Stimulus for rapid ocular dominance plasticity in visual cortex. *Journal of Neurophysiology, 95*, 2947–2950.

Roberts, EB; Ramoa, AS (1999). Enhanced NR2A subunit expression and decreased NMDA receptor decay time at the onset of ocular dominance plasticity in the ferret. *Journal of Neurophysiology, 81*, 2587–2591.

Rosenmund, C; Feltz, A; Westbrook, GL (1995). Calcium-dependent inactivation of synaptic NMDA receptors in hippocampal neurons. *Journal of Neurophysiology*, 73, 427–430.

Rubin, JE; Gerkin, RC; Bi, GQ; Chow, CC (2005). Calcium time course as a signal for spike-timing-dependent plasticity. *Journal of Neurophysiology, 93*, 2600–2613.

Sabatini, BL; Oertner, TG; Svoboda, K (2002). The life cycle of Ca^{2+} ions in dendritic spines. *Neuron, 33*, 439–452.

Schneggenburger, R (1996). Simultaneous measurement of Ca^{2+} influx and reversal potentials in recombinant N-methyl-D-aspartate receptor channels. *Biophysical Journal, 70*, 2165–2174.

Shah, NT; Yeung, LC; Cooper, LN; Cai, Y; Shouval, HZ (2006). A biophysical basis for the inter-spike interaction of spike-timing-dependent plasticity. *Biological Cybernetics*, 95, 113–121.

Shatz, CJ (1990). Impulse activity and the patterning of connections during CNS development. *Neuron, 5*, 745–756.

Shi, J; Aamodt, SM; Constantine-Paton, M (1997). Temporal correlations between functional and molecular changes in NMDA receptors and GABA neurotransmission in the superior colliculus. *Journal of Neuroscience, 17*, 6264–6276.

Shouval, HZ; Bear, MF; Cooper, LN (2002). A unified model of NMDA receptor-dependent bidirectional synaptic plasticity. *Proceedings of the National Academy of Sciences of the United States of America, 99*, 10831–10836.

Shpiro, A; Curtu, R; Rinzel, J; Rubin, N (2007). Dynamical characteristics common to neuronal competition models. *Journal of Neurophysiology, 97*, 462–473.

Sjöström, PJ; Nelson, SB (2002). Spike timing, calcium signals and synaptic plasticity. *Current Opinion in Neurobiology, 12*, 305–314.

Sjöström, PJ; Turrigiano, GG; Nelson, SB (2001). Rate, timing, and cooperativity jointly determine cortical synaptic plasticity. *Neuron, 32*, 1149–1164.

Song, S; Abbott, LF (2001). Cortical development and remapping through spike timing-dependent plasticity. *Neuron, 32*, 339–350.

Song, S; Miller, KD; Abbott, LF (2000). Competitive Hebbian learning through spike-timing-dependent synaptic plasticity. *Nature Neuroscience, 3*, 919–926.

Stephenson, FA (2001). Subunit characterization of NMDA receptors. *Current Drug Targets*, 2, 233–239.

Svoboda, K; Denk, W; Kleinfeld, D; Tank, DW (1997). In vivo dendritic calcium dynamics in neocortical pyramidal neurons. *Nature, 385*, 161–165.

Tanabe, S; Pakdaman, K (2001). Noise-enhanced neuronal reliability. *Physical Review E, 64*, 041904.

Taniike, N; Lu, YF; Tomizawa, K; Matsui, H (2008). Critical differences in magnitude and duration of N-methyl-D-aspartate (NMDA) receptor activation between long-term potentiation (LTP) and long-term depression (LTD) induction. *Acta Medica Okayama, 62*, 21–28.

Tegnér, J; Kepecs, Á (2002). Why neuronal dynamics should control synaptic learning rules. *Advances in Neural Information Processing Systems, 14*, 285–292.

Umemiya, M; Chen, N; Raymond, LA; Murphy, TH (2001). A calcium-dependent feedback mechanism participates in shaping single NMDA miniature EPSCs. *Journal of Neuroscience, 21*, 1–9.

Wang, XJ (1998). Calcium coding and adaptive temporal computation in cortical pyramidal neurons. *Journal of Neurophysiology, 79*, 1549–1566.

Wiesel, TN (1982). Postnatal development of the visual cortex and the influence of environment. *Nature, 299*, 583–591.

Yang, SN; Tang, YG; Zucker, RS (1999). Selective induction of LTP and LTD by postsynaptic [Ca^{2+}]i elevation. *Journal of Neurophysiology, 81*, 781–787.

Yeung, LC; Castellani, GC; Shouval, HZ (2004). Analysis of the intraspinal calcium dynamics and its implications for the plasticity of spiking neurons. *Physical Review E, 69*, 011907.

Zador, A; Koch, C; Brown, TH (1990). Biophysical model of a Hebbian synapse. *Proceedings of the National Academy of Sciences of the United States of America, 87*, 6718–6722.

Zhang, LI; Tao, HW; Holt, CE; Harris, WA; Poo, MM (1998). A critical window for cooperation and competition among developing retinotectal synapses. *Nature, 395*, 37–44.

In: Visual Cortex: Anatomy, Functions … ISBN: 978-1-62100-948-1
Editors: J.M. Harris et al. pp. 37-67 © 2012 Nova Science Publishers, Inc.

Chapter 2

ELECTROPHYSIOLOGICAL ASSESSMENT OF THE HUMAN VISUAL SYSTEM

***Takao Yamasaki**[*1,2] **and Shozo Tobimatsu**[1]*

[1] Department of Clinical Neurophysiology, Neurological Institute, Graduate School of Medical Sciences, Kyushu University, Fukuoka, Japan.
[2] Department of Neurology, Minkodo Minohara Hospital, Fukuoka, Japan.

ABSTRACT

In humans, visual information is processed simultaneously via multiple parallel channels. Condensed and parallel signals from the retina arrive in the primary visual cortex via the lateral geniculate nucleus. These signals then remain segregated until the higher levels of visual cortical processing through at least two separate but interacting parallel pathways; the ventral and dorsal streams. The former projects to the inferior temporal cortex for processing form and color, because it can detect visual stimuli with high spatial frequency and color. In contrast, the latter connects to the parietal cortex for detecting motion, because it responds to high temporal frequency stimuli. Based on these distinct physiological characteristics, we hypothesized that manipulating visual stimulus parameters would enable us to evaluate the different levels of each stream. So far, we have developed several techniques to record

[*] Corresponding author: Takao Yamasaki, M.D., Ph.D., Department of Clinical Neurophysiology, Neurological Institute, Graduate School of Medical Sciences, Kyushu University, 3-1-1 Maidashi, Higashi-ku, Fukuoka 812-8582, Japan, E-mail: yamasa@neurophy.med.kyushu-u.ac.jp, Tel: +81-92-642-5542; Fax: +81-92-642-5545.

visual evoked potentials (VEPs) and event-related potentials (ERPs) with optimal stimuli. In this review, we first summarize current concepts of the major parallel visual pathways. Second, we describe the relationship between the parallel visual pathways and higher visual system dysfunction. Third, we introduce VEP and ERP techniques that can assess the function of each stream and region of visual cortex. Finally, we address the clinical applications of VEP and ERP recording techniques for several neurological disorders involving specific visual dysfunction.

1. INTRODUCTION

The visual system is the dominant sensory system in humans. This system comprises the eyes, the connecting pathways through to the visual cortex and other parts of the brain. Approximately 25% of the human brain is involved in visual processing [1]. The visual system separates different types of information into anatomically segregated parallel streams of processing. In primates, the tracts from the retina to the primary visual cortex (V1) are clearly split into two pathways: the magnocellular (M) and parvocellular (P) pathways. These pathways differ markedly along several anatomical and physiological dimensions [1,2]. The higher levels of visual cortex also contain two streams of processing, each of which includes multiple visual areas and mediates different visual behaviors. The dorsal stream includes areas in the parietal cortex and is important for vision related to motion or spatial relationships. Conversely, the ventral stream includes visual areas in the temporal lobe and is more involved in the analysis of form and color [1,2].

Over the past few decades, the use of non-invasive techniques, including electrophysiology and neuroimaging, has dramatically increased our detailed knowledge of the functional organization of the human visual system. It is well acknowledged that visual evoked potentials (VEP) and event-related potentials (ERPs) are useful for investigating the physiology and pathophysiology of the human visual system, including visual pathways and the visual cortex. VEPs and ERPs can be used effectively in association with psychophysical techniques to study both normal and abnormal visual function [1]. VEPs and ERPs may detect abnormalities in patients with visual complaints but no objective findings on examination and in patients without visual symptoms [1]

In this review, we first summarize current concepts of the major parallel visual pathways. Second, we describe the relationship between the parallel visual pathways and higher visual dysfunction. Third, we introduce VEP and

ERP techniques that can assess the function of each stream and each region of visual cortex. Finally, we describe the clinical applications of VEPs and ERPs for several neurological disorders involving specific types of visual dysfunction.

2. Anatomy and Physiology of the Parallel Visual Pathways

2.1. The Parallel Visual Pathways

In humans, visual information from the retina projects to the V1 via the lateral geniculate nucleus (LGN) [1,2]. Ganglion cells are the output cells of the retina. Many types of ganglion cells exist, but midget, parasol and bistratified ganglion cells comprise approximately 90% of all ganglion cells [3]. These cells functionally complement each other to extend the range of vision in terms of wavelength and spatio-temporal frequency [3]. Midget ganglion cells are considered to constitute the origin of the P-pathway, and comprise approximately 70% of the total population of cells projecting to the LGN [4]. These cells convey a red/green color-opponent signal to the P-layers of the LGN, which in turn project to layer 4Cβ of V1 [4,5]. Cells in this pathway typically exhibit small receptive fields, low contrast sensitivity, slow axonal conduction velocities and sensitivity to high spatial and low temporal frequencies [1,2]. Parasol ganglion cells are considered to be the origin of the M-pathway and comprise approximately 10% of the total population of cells projecting to the LGN [4]. These cells convey a broadband, achromatic signal to the M-layers of the LGN and to layer 4Cα of V1 [4,5]. Cells in this pathway generally exhibit large receptive fields, high contrast sensitivity, fast axonal conduction velocities and sensitivity to high temporal and low spatial frequencies [1,2].

As mentioned above, two major cortical visual pathways have been proposed: the ventral or temporal and the dorsal or parietal streams [6,7]. The ventral stream is also commonly referred to as the 'what' system because it is involved in the identification of objects, whereas the dorsal stream is referred to as the 'where' system, because of its involvement in processing spatial location. The M and P pathways are thought to correspond approximately to the two systems, with the P pathway projecting primarily to the ventral stream and the M pathway providing the primary input to the dorsal stream [1,2]. The

P (what) system projects to area V4 via the P-blob and P-interblob pathways of V1, then proceeds to the inferior temporal (IT) area [1,2]. These areas are involved in the processing of visual form (processed in the P-interblob pathway) and color (processed in the P-blob pathway). Conversely, the M (where) system projects to area V3A via V1 and V2, then to V5/MT+ (V5/MT and MST) and V6, and terminates in the posterior parietal area [1,2,8]. This system processes the location of stimuli, and determines whether or not visual stimuli are moving. More recently, Rizzolatti and Matelli [8] proposed that two anatomically segregated subcircuits of the dorsal stream might mediate different behavioral goals. These authors speculated that the dorso-dorsal (d-d) pathway, proceeding through V6 and the superior parietal lobule (SPL), is concerned with the 'online' control of action (while the action is ongoing), whereas the ventro-dorsal (v-d) pathway, through V5 and the inferior parietal lobule (IPL), is concerned with spatial perception and 'action understanding' (the recognition of actions performed by others). Another series of studies demonstrated that macaque V6 is connected with visual areas including V1-3, V3A, V5/MT+, SPL and IPL [10,11]. A schematic representation of the general concept of parallel visual pathways is shown in Fig.1.

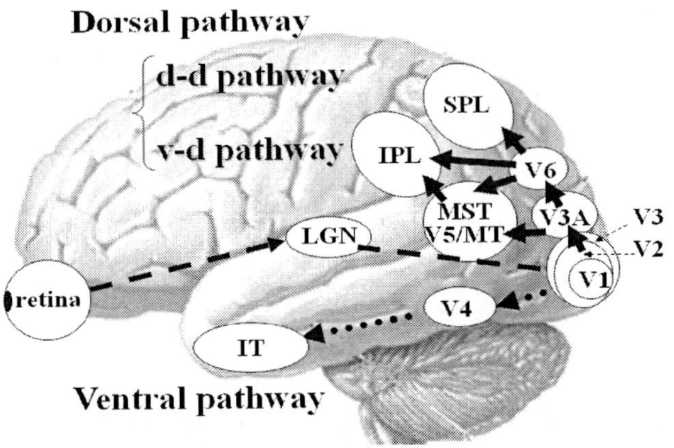

Figure 1. Parallel visual pathways in humans. Abbreviations in this and subsequent figures: d-d pathway, dorsodorsal pathway; v-d pathway, ventrodorsal pathway; LGN, lateral geniculate nucleus; V1, 2, 3, 4 and 6, primary, secondary, tertiary, quaternary and senary visual cortices; V3A, V3 accessory; V5/MT, quinary visual cortex/middle temporal area; MST, medial superior temporal area; IPL, inferior parietal lobule, SPL, superior parietal lobule; IT, inferior temporal cortex. (Adapted from Yamasaki et al., 2012 [65]).

2.2. Ventral Stream

2.2.1. P-Blob Pathway

The P-blob pathway is thought to be specialized for the analysis of color. Color processing begins with the absorption of light by cone photoreceptors. There are three types of cone photoreceptors, termed L-, M-, and S-cones, which are primarily sensitive to long wavelength light (560 nm), medium wavelength light (530 nm) and short wavelength light (420 nm) within the visible spectrum, respectively [9,12]. The cone signals are processed by several classes of retinal ganglion cells. Most retinal ganglion cells send their axons to the LGN [9,12]. In general, receptive fields in the LGN are very similar to those of the retinal ganglion cells that provide input to the LGN. Processing at the level of the ganglion cells and LGN is based on color opponency. Two types of cone-opponent cells have been described: those that compare L activation to M activation, referred to as red/green cells, and those that compare S activation to some combination of L and M activation, referred to as blue/yellow cells [9,12]. Such cone-opponent cells perform the type of calculations necessary to disambiguate wavelength and intensity, and provide the building blocks of color vision. LGN projections arrive in the V1 along anatomically segregated streams that keep red/green signals separate from blue/yellow signals.

In V1, there are three types of cells; luminance-preferring, color-luminance, and color-preferring cells [13,14]. Most cells classified as color-luminance cells are cone opponent, while most cells classified as luminance-preferring cells are cone non-opponent. All color-preferring cells are cone opponent. In particular, the color processing of V1 is characterized by double-opponent (both for color and space) color-luminance cells. Double-opponent neurons compare color signals across visual space. These neurons are so named because their receptive fields are both chromatically and spatially opponent. Double-opponent cells are candidates for the neural basis of color contrast and color constancy [9,12]. Color-tuned neurons in V1 coincide with cytochrome-oxidase blobs, which send projections specifically to cytochrome-oxidase thin-stripes of V2 [9,12]. “Globs” represent the next stage beyond V2 in the hierarchy of color processing. Glob cells elaborate the perception of hue. The term ‘globs’ draws an analogy to the cytochrome-oxidase blobs of V1 [15]. Globs are located within the posterior inferior temporal (PIT) cortex, a brain region that encompasses area V4, and brain regions immediately anterior to V4. Color signals are then processed by regions deep within the IT cortex. IT integrates color perception in the context of behavior. Finally, although the details of this process remain unclear, these signals appear to interface with motor programs and emotional centers of the brain to mediate the widely acknowledged emotional salience of color [9,12].

2.2.2. P-Interblob Pathway

Form information is mainly processed in the P-interblob ventral pathway. This pathway can be characterized by hierarchical architecture in which neurons in higher areas code for progressively more complex representations by pooling information from lower areas [16,17]. Hence, neurons in V1 code for relatively simple features such as local contours, whereas neurons in IT fire in response to whole complex forms [18]. In monkeys, many V1 cells function as local spatio-temporal filters, responsive to oriented bars [19]. V2 cells respond to illusory contours of figures [20]. Some V4 cells respond only if a stimulus has a specific color or pattern [21,22]. Occipito-temporal cells respond selectively to particular shapes [23-30].

Functional magnetic resonance imaging (fMRI) studies in humans have demonstrated that the ventral stream contains a small number of category-specific regions, which are primarily involved in processing specific stimulus classes [31,32]. Object-selective region exists in the lateral occipital cortex, occipitotemporal sulcus and fusiform gyrus (which together make up the lateral occipital complex; LOC). The LOC responds more strongly to objects than to scrambled object stimuli [33,34], and is involved in object recognition [35]. Face-selective regions include a region in the fusiform gyrus (the fusiform face area (FFA)) [36], a region in the inferior occipital gyrus (the occipital face area (OFA)) [32] and a region in the posterior superior temporal sulcus (STS) [32]. These regions respond more strongly to faces than to objects [36-39]. A region in the parahippocampal gyrus (the parahippocampal place area; PPA) [40] responds more to places than to faces or objects, and is involved in place perception [41] and memory [42]. The extrastriate body area (EBA) is involved in recognizing individuals by their posture and the perception of one's own body parts [43]. Letter strings and words are thought to be processed in a region in the left fusiform gyrus, known as the visual word form area (VWFA) [44,45].

2.3. Dorsal Stream

2.3.1. Ventro-Dorsal Pathway

The v-d pathway is important for motion and space perception and includes V5/MT and the IPL [8]. Human V5/MT+ is functionally and anatomically located in the depths of the anterior occipital sulcus (the ascending limb of the inferior temporal sulcus) and the anterior portions of either the inferior lateral occipital or the inferior occipital sulcus [47,48]. V5/MT has a high density of motion-sensitive neurons [49,50]. V1 neurons have small receptive fields that can detect local motion. The characteristic properties of V5/MT neurons include larger receptive fields than V1 cells [51,52], center-surround interactions [53], integration of different directions [54], and sensitivity to motion coherence [55]. Taken together, these findings suggest that V5/MT integrates local motion signals from V1 into the more global representations of motion needed as a basis for perceptual performance. Projections from V5/MT into the neighboring MST area and the parietal lobe appear to provide a good neural substrate for the use of visual motion in the control of eye movements and other actions. For these reasons, V5/MT is often considered a key region in extrastriate motion processing (lateral motion area [46]) [1].

In primates, neurons in area 7a and the ventral intraparietal area correspond to the human IPL [8,56], and preferentially respond to expanding optic flow (OF) stimulation [57,58]. Radial OF is the visual motion seen during observer self-movement, and is known to be important in daily life because it provides cues about heading direction and the three-dimensional structure of the visual environment [59,60]. fMRI studies on humans have reported that the IPL plays a crucial role in high-level motion perception [61,62] including radial OF perception [63-65]. Furthermore, the analysis performed in the IPL involves the integration of visual, auditory and somatosensory stimuli for action on and perception of the external world [56].

2.3.2. Dorso-Dorsal Pathway

The d-d pathway also plays an important role in motion perception and consists of V6 and the SPL [8]. Recent results suggest that V6 should be considered an additional medial motion processing area, as well as a lateral motion processing area (V5/MT) [10,46]. V6 is located in the parieto-occipital sulcus of macaques and humans. In macaques, V6 abuts the end (the representation of the far periphery) of V3 and V3A. It has a clear retinotopic organization, representing the contralateral hemifield. Most of its cells are visually responsive, and approximately 75% are direction sensitive. Similar to V5/MT+ cells, the receptive field of cells in V6 is much larger than that of cells in V1. The adjacent area, V6A, which occupies the dorsal/anterior portion of the sulcus, has no obvious retinotopic organization and only around 60% of the neurons are visually responsive. Visual neurons in this area are again predominantly motion sensitive. These findings suggest that macaque V6 and V6A play a pivotal role in providing visual motion information to the motor system [66-69]. Similarly to monkey V6, human V6 is confined to the dorsal portion of the parieto-occipital sulcus, occupying the fundus and posterior bank of the sulcus. This area contains a complete representation of the contralateral hemifield, with the lower field located medially and more anterior to V3/V3A, extending dorsally to the upper field. As in non-human primates, human V6 responds more to coherent than incoherent motion [46].

A key feature of neurons in the SPL is their ability to combine different neural signals relating visual target location, eye and/or hand position and movement direction into coherent frame of spatial reference [70]. Human fMRI studies have reported that the SPL is highly responsive to unidirectional coherent motion stimuli [71]. A previous fMRI study in our laboratory demonstrated that the SPL is significantly activated by horizontal motion (HO), but not by radial OF [63-65]. Therefore, it is likely that SPL processes simple unidirectional motion information.

3. The Relationship between the Parallel Visual Pathways and Higher-Level Visual Dysfunctions

3.1. Dysfunction of the Ventral Stream

A number of clinical syndromes are related to lesions of occipitotemporal regions within the ventral stream. Most of these syndromes are manifested as a distinct form of visual agnosia. Visual agnosia is defined as the impaired recognition of objects, which is not caused by a sensory deficit or generalized intellectual loss [72]. Visual agnosia is commonly divided into two types: apperceptive and associative agnosia. Patients suffering from apperceptive agnosia are unable to perceive an object because visual information is not integrated into a global percept, leading to “piecemeal perception”. Apperceptive agnosia is usually associated with posterior brain lesions. In contrast, patients with associative agnosia tend to possess intact global perception of visual information and are able to match and copy shapes, but are unable to identify objects or categories of objects [73]. Associative agnosia is thought to result from either occipitotemporal lesions that cause a disconnection between areas responsible for basic visual perception and memory systems involved in object recognition, or from a deficit of the perception of a specific category of visual forms, such as faces, objects or words [74].

One specific type of agnosia is prosopagnosia. Patients with prosopagnosia cannot recognize familiar faces or learn new faces [75]. Impaired recognition is specifically limited to faces, since patients are still able to recognize their families and friends, relying on cues from posture, voice or clothing. Most patients with prosopagnosia suffer from bilateral damage to the inferior occipitotemporal cortex, in particular the lingual and fusiform gyri [75,76], but there are also numerous reports of prosopagnosia after unilateral right-hemisphere lesions [77].

Another type of agnosia, limited to reading disabilities, is pure alexia. Such patients have severely impaired recognition of single-word forms, but are still able to read letter-by-letter [78]. These deficits are attributed to damage to a specific area located in the left mid fusiform gyrus (VWFA), or disconnection between V1 and the VWFA.

Cerebral achromatopsia is an acquired deficit in color perception caused by damage to the ventromedial cortex, particularly bilateral lingual gyri or fusiform gyri [79,80]. Affected patients describe a world that looks faded, gray, and washed out, or completely devoid of color, like a black-and-white photograph. This condition is often associated with a superior homonymous visual field defect from damage to the inferior striate cortex [81]. In addition, cerebral achromatopsia is often accompanied by visual object agnosia, prosopagnosia, pure alexia, topographagnosia and defects of visual memory [82].

3.2. Dysfunction of the Dorsal Stream

Several syndromes are linked to lesions in the dorsal stream. One of the rarest syndromes is akinetopsia, which refers to the loss of perception of visual motion with other visual functions, such as perception of form and color being intact [83]. Akinetopsia is thought to be caused by bilateral damage of area V5/MT [84,85], while unilateral damage to area V5/MT may lead to more subtle deficits in motion processing or hemiakinetopsia [86]. Unilateral lesions of V6 can lead to direction-selective motion blindness, similar to that caused by damage to V5/MT [87].

Bálint's syndrome is also associated with a specific set of symptoms related to impaired visual processing in the dorsal stream. Bálint's syndrome is classically defined as a triad of simultanagnosia, optic ataxia and oculomotor apraxia [88]. Patients with simultanagnosia are unable to identify two items presented simultaneously because they cannot integrate visual information into a global representation of space [73]. Simultanagnosia is most commonly caused by bilateral occipitoparietal damage [89]. In optic ataxia, visual guidance of movements towards objects is impaired due to a disconnection between cortical motor systems and visual inputs [90]. In oculomotor apraxia, patients are impaired at making voluntary eye movements from one fixation point to another fixation point in space.

Probably the best-known disorder related to impaired dorsal processing is unilateral neglect, which can refer to a failure, difficulty or slowness in reporting, interacting with or moving towards objects, sounds or representations as a consequence of their spatial position, most frequently occurring in the left visual field. Although both left and right hemisphere stroke patients may suffer from neglect in the acute phase [91], chronic neglect is almost exclusively caused by right hemisphere lesions [92,93]. One of the striking aspects of neglect is that it can be caused by a variety of brain regions, which are all part of a complex network that is involved in the modulation of spatial attention. These brain regions include the inferior parietal lobe, the middle temporal lobe, the superior temporal lobe and subcortical areas such as the basal ganglia and the thalamus [94].

4. Electrophysiological Techniques to Evaluate Parallel Processing

4.1. VEP Technique for Testing Lower-Level Ventral (P-) and Dorsal (M-) Pathways

VEPs are useful in evaluating the function of the lower-level visual cortex. Any repetitive visual stimulus can be used to elicit VEPs. However, selecting an appropriate stimulus is very important. Pattern stimuli (gratings and checks) are preferential for exploring the function of V1 because local spatial frequency analyzers are presumably present in V1 [95,96]. Pattern stimuli are defined by the type of pattern (grating, bars or checks), spatial frequency, contrast, field size, mean luminance of the stimulus and background, the type of pattern presentation (reversal or onset-offset), and the temporal frequency of the reversal or the presentation [1,97]. The most commonly used method of stimulation is pattern reversal using either a checkerboard or a grating pattern [98]. Pattern reversal-VEPs are characterized by an initial small negative potential (N75), followed by a major positive wave (P100), then another negative wave (N145) [1,97]. Magnetoencephalography (MEG) studies have suggested that P100m (corresponding to the P100) originates in the occipital cortex (V1) [99-101].

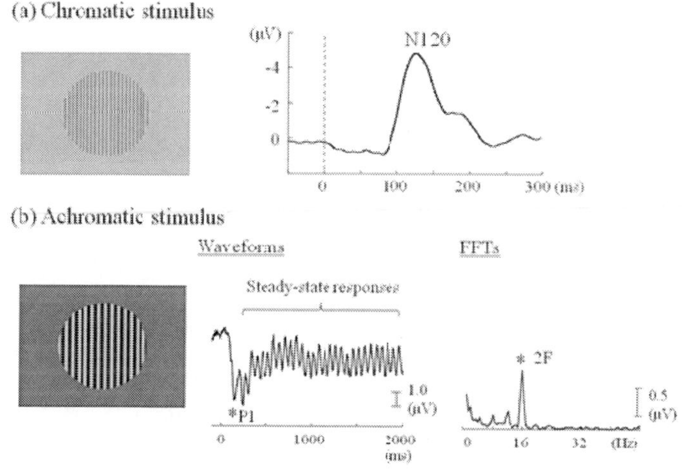

Figure 2. VEPs with chromatic and achromatic stimuli. (a) In chromatic VEPs, red/green chromatic sinusoidal gratings with equal luminance (visual angle, 10×10°; mean luminance, 21 cd/m^2; spatial frequency, 2 cycles per degree) were presented for 200 ms. A negative component (N120) was elicited in occipital regions (maximum at Oz electrode). (b) In achromatic VEPs, achromatic (black/white) low-contrast (16.6 %) sinusoidal gratings (visual angle, 10×10°; mean luminance, 21 cd/m^2; spatial frequency, 1 cycle per degree) were presented for 2,000 ms and rapidly alternated in a square-wave fashion at 8 Hz (16 reversals/s). A positive component at around 120 ms (P1) and subsequent quasi-sinusoidal waveforms corresponding to the reversal frequency (16 Hz) were elicited at the occipital regions (maximum at Oz electrode). In the fast Fourier transform spectra (FFTs), the second harmonic (2F) component was found to be a major component.

As previously mentioned, the P- and M-pathways have distinct physiological characteristics. Thus, by manipulating these stimulus parameters, we can explore the function of P- and M-pathways in V1 in more detail. Two forms of VEP exist; transient and steady-state VEPs [1]. Transient VEPs are obtained to low stimulus rates, while steady-state VEPs are elicited by repetitive frequent stimuli [1]. Transient VEPs at low temporal frequencies elicited by chromatic (red-green) sinusoidal gratings with equal luminance and high spatial frequency are suitable for examining the P-pathway at the lower levels within V1 (Fig. 2a) [102-106]. This stimulus evokes a characteristic negative wave (N120) with a peak latency around 120 ms (Fig. 2a). Conversely, steady-state VEPs at high temporal frequencies that use achromatic (black-white) sinusoidal gratings with low contrast and low spatial frequencies are useful for evaluating the M-pathway within V1 (Fig. 2b). This stimulation induces a positive peak (P1) at around 120 ms followed by steady-state responses (Fig. 2b) [106,107].

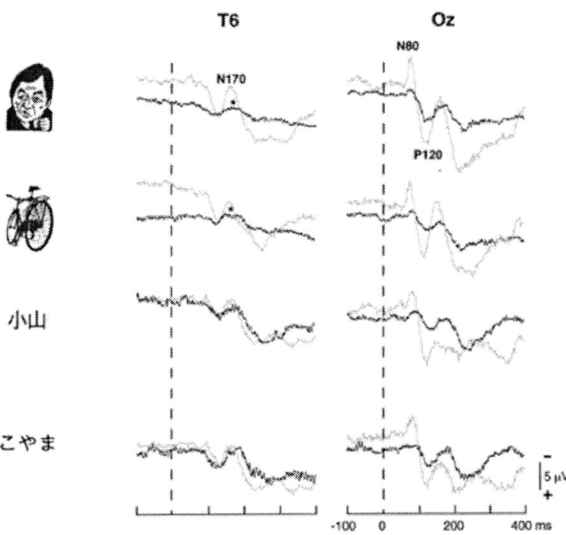

Figure 3. Category-specific ERPs in response to realistic drawing of a face (former prime minister), object (bicycle), and words (Chinese characters and Japanese syllabary). Thick traces indicate the patient's data while thin traces represent average data of normal subjects (n = 15). With face and object stimuli, the P190 was not obtained at Cz and the amplitude of the N170 at T6 was low (asterisks). With word stimuli, P190 at Cz was evoked with low amplitude (arrows) while the N170 at appeared normal at T6. For all stimuli, primary responses at Oz (N80/P120) were evoked normally with low amplitude. (Adapted from Yamasaki et al., 2004 [108]).

4.2. ERP Technique for Testing Higher-Level Ventral Pathway

Investigating the function of the entire visual system, including higher-level integrating processes, is difficult using VEPs alone. However, ERPs have advantage of studying covert higher-level neural processes, providing a non-invasive real-time measure. In the case of higher-level P-interblob function, category-specific ERPs using face, object and word stimuli have been successfully used to evaluate the functional specialization of the IT cortex (Fig. 3) [108]. ERPs have shown that objects categories can be reliably distinguished by the associated electrophysiological activity recorded on the surface of the occipito-temporal cortex (around T5 and T6). The largest and most consistent ERP/MEG difference has been observed as early as 130-170 ms, as the difference between pictures of faces and other objects at occipito-temporal sites [109-116]. More precisely, the first negative occipito-temporal component evoked by these stimuli is usually referred to as the N170 (Fig. 3). The N170 was right lateralized for faces [117-120], smaller and bilateral for objects [118], and as large for printed words in the left hemisphere as for faces [121-125]. These ERP findings are consistent with the results of neuropsychological and neuroimaging studies (showing a hemispheric advantage for the processing of words [left], faces [right] and objects [bilateral]), indicating a major functional distinction within the human object recognition system.

Regarding higher-level P-blob function, two studies recorded ERPs in a color discrimination task (Fig. 4a) [108,126]. In this task, 128 dots were presented in a random spatial pattern on a uniform blue background, with an onset-offset mode of presentation. Red dots were used as frequent stimuli (non-targets), while green stimuli were used as infrequent stimuli (targets). Infrequent stimuli were presented randomly, between presentations of frequent stimuli. The frequent/rare stimulus ratio in this task was 80:20. The results of these studies indicated that the P400 (P) was evoked at the Pz electrode [108,126] (Fig. 4b). However, the P400 (P) was not recorded in a subject with a color vision deficit affecting their perception of green [126], suggesting that the P400 (P) reflects the function of the P-pathway.

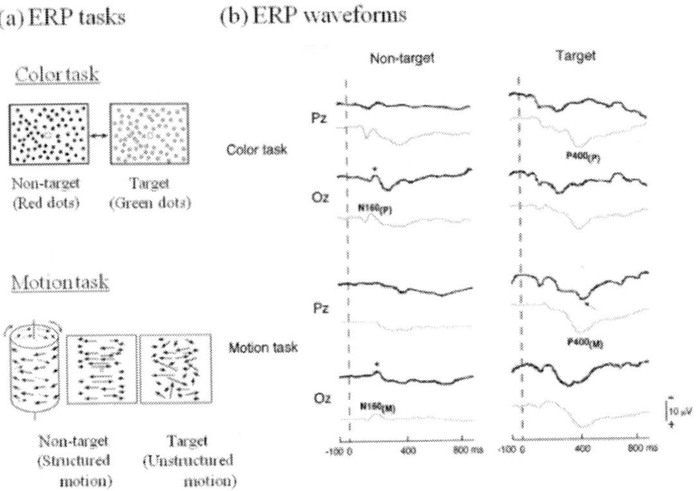

Figure 4. ERPs in color and motion discrimination tasks. (a) In the color task, red (frequent) or green (infrequent) dots were randomly presented. In the motion task, 3D structured (frequent) or unstructured (infrequent) motion was randomly presented. In both tasks, the frequent/rare stimulus ratio was 80:20. (b) ERP waveforms under the color and motion discrimination tasks. Thick traces indicate the patient's data while thin traces represent averaged data of normal subjects (n = 11). In the color task, P400 (P) at Pz was not evoked while N160 (P) at Oz was normal. In the motion task, P400 (M) at Pz (arrow) and N160 (M) at Oz were normal. (Adapted from Yamasaki et al., 2004 [108]).

4.3. ERP Technique for Testing Higher Level of the Dorsal Pathway

Motion stimuli can be useful for examining higher-level dorsal function. Visual motion stimuli can be characterized by the direction of motion: linear translation, rotation, expansion and contraction, and motion in depth. One particularly useful stimulus is the random dot kinematogram, in which the overall motion is extracted from a set of coherently and/or incoherently moving subunits. A distinction should also be made among real motion, apparent motion (stepwise dislocation) and illusory motion (motion aftereffects, etc.) [1]. Coherent motion stimuli using random dots have been widely used in psychophysical, electrophysiological, and neuroimaging studies to investigate global motion processing [127-129]. Coherent global motion, such as radial OF and HO motion, have been used in a number of studies (Fig.

5a). Radial OF is a type of complex visual motion related to self-motion perception [59,60], while HO refers to simple unidirectional motion. As mentioned earlier, the dorsal stream is divided into two functional streams, the v-d stream and the d-d stream [8]. The d-d stream (SPL) is more closely related to HO motion processing, while the v-d stream (IPL) is important for OF motion processing [63-65]. ERP studies reported that perception of these stimuli was associated with two major components (N170, P200) [63-65]. The occipitotemporal N170 exhibited a V5/MT origin, and was evoked by both types of stimuli. In contrast, the parietal P200 was found to originate in IPL (BA 40), and was only elicited by OF stimuli (Fig. 5b) [63-65].

(a) Coherent motion stimuli

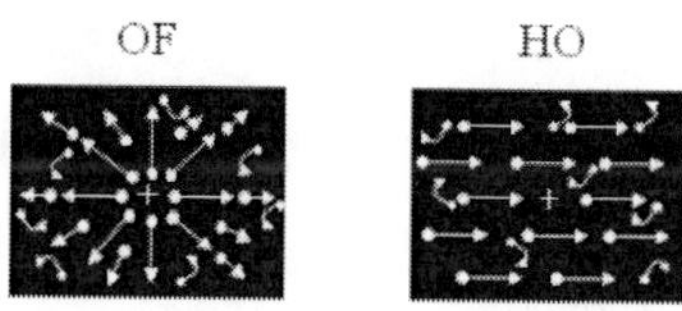

(b) ERP waveforms

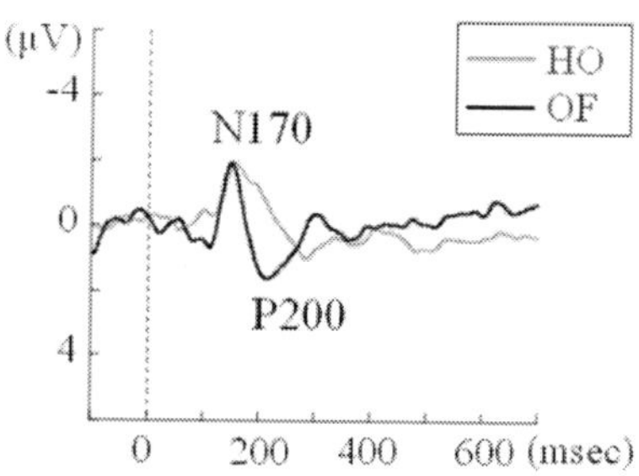

Figure 5. (a) Visual motion stimuli. Four hundred white square dots (visual angle, 0.2×0.2°; luminance, 48 cd/m^2) are randomly presented on a black background (visual angle, 50×48°; luminance, 0.1 cd/m^2). The contrast level is 99.6%. The white dots moved at a velocity of 5.0°/s. When the white dots moved incoherently, random stimulation was created. When the white dots moved coherently, radial OF and HO motion were perceived. (b) ERP waveforms in healthy subjects. Two major components (N170, P200) were obtained. N170 was evoked by both HO and OF stimuli while P200 was only elicited by OF. Abbreviations in this and subsequent figures: optic flow, OF; horizontal motion, HO (Adapted from Yamasaki et al., 2012 [65]).

ERP measurement in conjunction with the motion discrimination task has also been found useful for investigating higher-level dorsal stream function (Fig. 4a) [108,126]. In this task, 3-D structured coherent (frequent) and unstructured incoherent (rare) motion stimuli are used. Rare stimuli are presented randomly between frequent stimuli. The frequent/rare stimulus ratio is 80:20. Consequently, P400 (M) is evoked at the Pz electrode (Fig. 4b). Interestingly, the P400 (M) was elicited in a subject with a deficit in green color vision [126], suggesting that it depends on the function of the M-pathway but not the P-pathway.

5. Clinical Applications of VEPs and ERPs

5.1. Dysfunction of the Ventral Stream

We reported a unique case of multiple sclerosis with associative visual agnosia, prosopagnosia and cerebral achromatopsia, but no alexia and no akinetopsia [108]. The patient was alert and fully oriented with normal intelligence, and his corrected visual acuity was sufficiently preserved. MRI revealed white mater lesions in ventral parts of the temporo-occipital lobes including the fusiform and lingual gyri and inferior longitudinal fasciculi. Fluorodeoxyglucose-positron emission tomography also demonstrated decreased glucose metabolism in the ventral parts of the temporo-occipital lobes, with right dominance. The patient was then examined with VEP and ERP techniques. To investigate lower-level visual function, pattern-reversal VEPs and isoluminant red/green VEPs were measured. Pattern-reversal VEPs to a black-white checkerboard pattern revealed normal P100 latency. Chromatic VEPs to isoluminant red-green sinusoidal gratings also demonstrated a normal N120 latency. These findings suggested that the patient exhibited preserved V1 function. Category-specific ERPs with faces, objects and words and ERPs were then measured in a color discrimination task to examine the patient's higher-level ventral function. In category-specific ERPs, the patient's N170 amplitude in response to faces and objects was significantly lower than that of normal controls, while N170 responses to words were normal (Fig. 3). The P400 (P) was not evoked in ERPs in the color discrimination task (Fig. 4b), while the P400 (M) was normal in ERPs in the motion discrimination task (Fig. 4b). It is likely that our patient had a selective impairment of subdivisions (objects, faces and color) within higher-level ventral stream processing. Thus, our electrophysiological findings were consistent with neuropsychological findings, and the results of neuroimaging examinations.

5.2. Dysfunction of the Dorsal Stream

Disturbance of motion perception is caused by parietal lobe dysfunction in Alzheimer's disease (AD) [130] as well as in stroke and brain trauma. AD is the most common form of dementia. In addition to impairment of episodic memory, higher visual dysfunction is one of the cognitive hallmarks of AD [131]. Various visual functions, including the perception of objects, faces, words and visuospatial stimuli are impaired in AD. However, deficits of visuospatial perception (i.e. the perception of space and motion) are the most prominent features. Such deficits play a critical role in the navigational impairments found in AD [132-134]. Some patients with mild cognitive impairment (MCI), the prodromal stage of AD, also exhibit impaired motion perception [124]. We recently examined the way in which the parallel visual pathways are functionally altered in MCI patients, using multimodal visual stimuli [63-65,135-138]. As a measure of lower-level visual function (V1), the P120 for chromatic stimulus and steady-state responses for achromatic stimulus were found to be normal. For higher-level dorsal stream function, we found no significant differences in N170 responses for both OF and HO stimuli between MCI patients and healthy controls. However, a significantly prolonged P200 latency for OF stimuli was observed in MCI patients compared to healthy controls. Thus, within the dorsal stream, the v-d stream (IPL) related to OF perception but not the d-d stream (SPL) was selectively impaired in MCI patients [63,65]. Conversely, ERPs in response to faces and Kanji-words (ideograms) were normal in MCI patients [135], suggesting that the ventral stream was intact. Overall, MCI patients exhibited a selective impairment of higher-level processing in the dorsal stream (v-d stream), consistent with the behavioral features of AD and MCI patients.

Autism spectrum disorder (ASD) is a neurodevelopmental disorder characterized by social interaction and communication impairments, as well as restricted and repetitive behaviors and interests [139]. Individuals with ASD exhibit superior performance in processing fine details [140-142], while even those with high IQ are poor at processing global structure and perceiving motion [143-145]. It was recently proposed that low-level perception contributes to higher-level impairments of social cognition in ASD [146,147]. In terms of the concept of parallel visual processing, atypical visual characteristics of superior processing of fine detail (local structure), inferior processing of global structure, and impaired motion perception in ASD may result from the hyperfunction of the ventral pathway and dysfunction of the dorsal pathway. To date, three distinct hypotheses have been proposed based on recently accumulating evidence regarding early processing of the visual system in ASD; the "pathway-specific" hypothesis [143], the "complexity-specific (or integration)" hypothesis [145,148] and the "M-pathway" hypothesis [149]. The pathway-specific hypothesis suggests dysfunction of the dorsal pathway with sparing of the ventral pathway. The complexity-specific hypothesis indicates impairment of neuro-integrative processing at higher cortical levels including both the ventral and dorsal pathways. The M-pathway hypothesis implies dysfunction of the lower-level subcortical dorsal pathway. We recently examined the function of the dorsal stream in ASD adults using VEPs with achromatic stimuli [106] and ERPs with coherent OF and HO stimuli [150]. ASD adults exhibited significantly prolonged N170 and P200 latencies for OF, but not HO stimuli (Fig. 6). In contrast, both P1 and steady-state responses were normal in ASD [106].

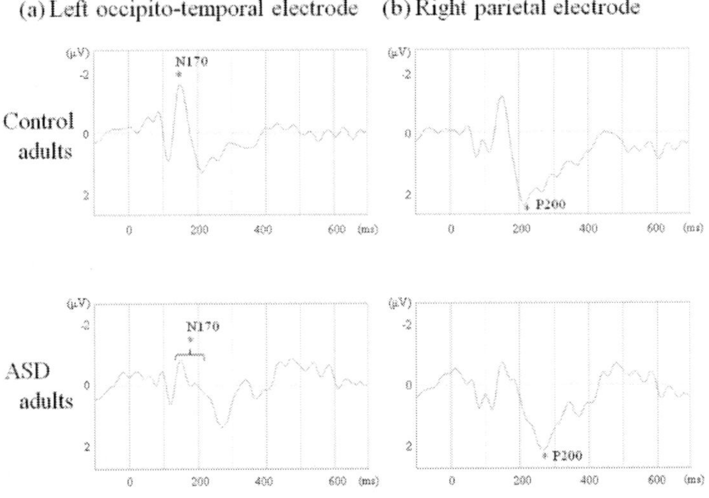

Figure 6. ERPs in response to OF stimulation at the left occipito-temporal electrode (a) and at the right parietal electrode (b) in control and ASD groups. ASD adults exhibited small and ill-defined N170 response (a) and prolonged P200 latency (b) compared with control adults. Abbreviations: autism spectrum disorders, ASD (Adapted from Yamasaki et al., 2011 [150]).

Therefore, these findings suggest impaired higher-level function (v-d stream) with preserved lower-level function of the dorsal stream [64,106,150]. These findings appear to provide partial support for the "complexity-specific" hypothesis over the "pathway-specific" or "M-pathway" hypotheses.

6. Conclusion

In humans, there are two major functional visual streams; the ventral and dorsal streams. Based on the distinct physiological characteristics of the two streams, we developed VEP and ERP recording techniques to evaluate the function of the ventral and dorsal streams at lower and higher levels, separately, by manipulating visual stimulus parameters. In doing so, we were able to detect specific visual impairments in various neurological disorders electrophysiologically, producing results that were consistent with the findings of clinical and neuroimaging examinations. These findings indicate that electrophysiological examination may be useful for evaluating the parallel visual pathways in various neurological disorders.

Acknowledgments

This study was supported in part by Grants-in-Aid for Scientists, No. 18890131 and No. 20591026, from the Ministry of Education, Culture, Sports, Science and Technology in Japan. This work was also supported by a grant from JST, RISTEX. We would like to thank Drs. Jun-ichi Kira, Yasumasa Ohyagi, Yoshinobu Goto, Takayuki Taniwaki, Akira Monji, Shigenobu Kanba, Takako Fujita, Yoko Kamio, Shinji Munetsuna, Takashi Yoshiura, Katsuya Ogata, Kensuke Sasaki, Motozumi Minohara, Katsuko Minohara, Ikue Ijichi and Yuka Miyanaga, Chiharu Kurita and Sachiko Takashima for their research assistance.

References

[1] Tobimatsu, S., & Celesia, G. G. (2006). Studies of human visual pathophysiology with visual evoked potentials. *Clinical Neurophysiology*, 117, 1414-1433.

[2] Livingstone, M., & Hubel, D. (1988). Segregation of form, color, movement, and depth: anatomy, physiology, and perception. *Science*, 240, 740-749.

[3] Schiller, P. H., & Logothetis, N. K. (1990). The color-opponent and broad-band channels of the primate visual system. *Trends in Neurosciences*, 13, 392-398.

[4] Dacey, D. M. (2000). Parallel pathways for spectral coding in primate retina. *Annual Review of Neuroscience*, 23, 743-775.

[5] Chatterjee, S., & Callaway, E. M. (2003). Parallel colour-opponent pathways to primary visual cortex. *Nature*, 426, 668-671.

[6] Mishkin, M., Ungerleider, L. G., & Macko, K. A. (1983). Object vision and spatial vision: two cortical pathways. *Trends in Neurosciences*, 6, 414-417.

[7] Tootell, R. B. H., Dale, A. M., Sereno, M. I., & Malach, R. (1996). New images from human visual cortex. *Trends in Neurosciences*, 19, 481-489.

[8] Rizzolatti, G., & Matelli, M. (2003). Two different streams form the dorsal visual system: anatomy and functions. *Experimental Brain Research*, 153, 146-157.

[9] Conway, B. R. (2009). Color vision, cones, and color-coding in the cortex. *The Neuroscientist*, 15, 274-290.

[10] Fattori, P., Pitzalis, S., & Galletti, C. (2009). The cortical visual area V6 in macaque and human brains. *Journal of Physiology, Paris*, 103, 88-97.

[11] Gamberini, M., Passarelli, L., Fattori, P., Zucchelli, M., Bakola, S., Luppino, G., & Galletti, C. (2009). Cortical connections of the visuomotor parietooccipital area V6Ad of the macaque monkey. *The Journal of Comparative Neurology*, 513, 622-642.

[12] Conway, B. R., Chatterjee, S., Field, G. D., Horwitz, G. D., Johnson, E. N., Koida, K., & Mancuso, K. (2010) Advances in color science: from retina to behavior. *The Journal of Neuroscience*, 30, 14955-14963.

[13] Johnson, E. N., Hawken, M. J., & Shapley, R. (2001) The spatial transformation of color in the primary visual cortex of the macaque monkey. *Nature Neuroscience*, 4, 409-416.

[14] Johnson, E. N., Hawken, M. J., & Shapley, R. (2004) Cone inputs in macaque primary visual cortex. *Journal of Neurophysiology*, 91, 2501-2514.

[15] Conway, B. R., Moeller, S., & Tsao, D. Y. (2007) Specialized color modules in macaque extrastriate cortex. *Neuron*, 56, 560-573.

[16] Logothetis, N. K., & Sheinberg, D. L. (1996) Visual object recognition. *Annual Review of Neuroscience*, 19, 577-621.

[17] Rolls, E. T. (2000) Functions of the primate temporal lobe cortical visual areas in invariant visual object and face recognition. *Neuron*, 27, 205-218.

[18] Rousselet, G. A., Thorpe, S. J., & Fabre-Thorpe, M. (2004) How parallel is visual processing in the ventral pathway? *Trends in Cognitive Sciences*, 8, 363-370.

[19] Hubel, D. H., & Wiesel, T. N. (1968). Receptive fields and functional architecture of monkey striate cortex. *The Journal of Physiology*, 195, 215-243.

[20] Peterhans, E., & von der Heydt, R. (1991). Subjective contours-bridging the gap between psychophysics and physiology. *Trends in Neurosciences*, 14, 112-119.

[21] Gallant, J. L., Braun, J., & Van Essen, D. C. (1993). Selectivity for polar, hyperbolic, and Cartesian gratings in macaque visual cortex. *Science,* 259, 100-103.

[22] Schein, S. J., & Desimone, R. (1990). Spectral properties of V4 neurons in the macaque. *The Journal of Neuroscience*, 10, 3369-3389.

[23] Booth, M. C., & Rolls, E. T. (1998). View-invariant representations of familiar objects by neurons in the inferior temporal visual cortex. *Cerebral Cortex,* 8, 510-523.

[24] Desimone, R., Albright, T. D., Gross, C. G., & Bruce, C. (1984). Stimulus-selective properties of inferior temporal neurons in the macaque. *The Journal of Neuroscience*, 4, 2051-2062.

[25] Fujita, I., Tanaka, K., Ito, M., & Cheng, K. (1992). Columns for visual features of objects in monkey inferotemporal cortex. *Nature*, 360, 343-346.

[26] Kobatake, E., & Tanaka, K. (1994). Neuronal selectivities to complex object features in the ventral visual pathway of the macaque cerebral cortex. *Journal of Neurophyiology*, 71, 856-867.

[27] Logothetis, N. K., Pauls, J., & Poggio, T. (1995). Shape representation in the inferior temporal cortex of monkeys. *Current Biology,* 5, 552-563.

[28] Sigala, N., & Logothetis, N. K. (2002). Visual categorization shapes feature selectivity in the primate temporal cortex. *Nature,*415, 318-320.

[29] Tanaka, K. (1993). Neuronal mechanisms of object recognition. *Science,* 262, 685-688.

[30] Wachsmuth, E., Oram, M. W., & Perrett, D.I. (1994). Recognition of objects and their component parts: responses of single units in the temporal cortex of the macaque. *Cerebral Cortex,* 4, 509-522.

[31] Kanwisher, N. (2003). The ventral visual object pathway in humans: evidence from fMRI. In: L. Chalupa, & J. Werner (Eds.), *The Visual Neurosciences*, (pp. 1179-1189). Cambridge, Massachusetts: MIT Press.

[32] Grill-Spector, K., Golarai, G., & Gabrieli, J. (2008). Developmental neuroimaging of the human ventral visual cortex. *Trends in Cognitive Sciences*, 12, 152-162.

[33] Grill-Spector, K., Kourtzi Z., & Kanwisher, N. (2001). The lateral occipital complex and its role in object recognition. *Vision Research*, 41, 1409-1422.

[34] Malach, R., Reppas, J. B., Benson, R. R., Kwong, K. K., Jiang, H., Kennedy, W. A., Ledden, P. J., Brady, T. J., Rosen, B. R., & Tootell, R. B. (1995). Object-related activity revealed by functional magnetic resonance imaging in human occipital cortex. *Proceedings of the National Academy of Sciences of the United States of America*, 92, 8135-8139.

[35] Grill-Spector, K., Kushnir, T., Hendler, T., & Malach, R. (2000). The dynamics of object-selective activation correlate with recognition performance in humans. *Nature Neuroscience*, 3, 837-843.

[36] Kanwisher, N., McDermott, J., & Chun, M. M. (1997). The fusiform face area: a module in human extrastriate cortex specialized for face perception. *The Journal of Neuroscience*, 17, 4302-4311.

[37] Halgren, E., Dale, A. M., Sereno, M. I., Tootell, R. B., Marinkovic, K., & Rosen, B. R. (1999). Location of human face-selective cortex with respect to retinotopic areas. *Human Brain Mapping*, 7, 29-37.

[38] Haxby, J. V., Ungerleider, L. G., Clark, V. P., Schouten, J. L., Hoffman, E. A., & Martin, A. (1999). The effect of face inversion on activity in human neural systems for face and object perception. *Neuron*, 22, 189-199.

[39] McCarthy, G., Puce, A., Gore, J. C., & Allison, T. (1997). Face-specific processing in the human fusiform gyrus. *Journal of Cognitive Neuroscience*, 9, 605-610.

[40] Epstein, R., & Kanwisher, N. (1998). A cortical representation of the local visual environment. *Nature*, 392, 598-601.

[41] Tong, F., Nakayama, K., Vaughan, J. T., & Kanwisher, N. (1998). Binocular rivalry and visual awareness in human extrastriate cortex. *Neuron*, 21, 753-759.

[42] Brewer, J. B., Zhao, Z., Desmond, J. E., Glover, G. H., & Gabrieli, J. D. (1998). Making memories: brain activity that predicts how well visual experience will be remembered. *Science*, 281, 1185-1187.

[43] Downing, P. E., Jiang, Y. H., Shuman, M., & Kanwisher, N. (2001). A cortical area selective for visual processing of the human body. *Science*, 293, 2470-2473.

[44] Dehaene, S., & Cohen, L. (2011). The unique role of the visual word form area in reading. *Trends in Cognitive Sciences*, 15, 254-262.

[45] Reich, L., Szwed, M., Cohen, L., & Amedi, A. (2011). A ventral visual stream reading center independent of visual experience. *Current Biology*, 21, 363-368.

[46] Pitzalis, S., Sereno, M. I., Committeri, G., Fattori, P., Galati, G., Patria, F., & Galletti, C. (2010). Human V6: the medial motion area. *Cerebral Cortex*, 20, 411-424.

[47] Wilms, M., Eickhoff, S. B., Specht, K., Amunts, K., Shah, N. J., Malikovic, A., & Fink, G. R. (2005). Human V5/MT+: comparison of functional and cytoarchitectonic data. *Anatomy and Embryology*, 210, 485-495.

[48] Malikovic, A., Amunts, K., Schleicher, A., Mohlberg, H., Eickhoff, S. B. Wilms, M., Palomero-Gallagher, N., Armstrong, E., & Zilles, K. (2007). Cytoarchitectonic analysis of the human extrastriate cortex in the region of V5/MT+: a probabilistic, stereotaxic map of area hOc5. *Cerebral Cortex*, 17, 562-574.

[49] Maunsell, J. H., & Newsome, W. T. (1987). Visual processing in monkey extrastriate cortex. *Annual Review of Neuroscience*, 10, 363-401.

[50] Zeki, S. M. (1974). Functional organization of a visual area in the posterior bank of the superior temporal sulcus of the rhesus monkey. *The Journal of Physiology*, 236, 549-573.

[51] Mikami, A., Newsome, W. T., & Wurtz, R. H. (1986). Motion selectivity in macaque visual cortex. I. Mechanisms of direction and speed selectivity in extrastriate area MT. *Journal of Neurophyiology*, 55, 1308-1327.

[52] Mikami, A., Newsome, W. T., & Wurtz, R. H. (1986). Motion selectivity in macaque visual cortex. II. Spatiotemporal range of directional interactions in MT and V1. *Journal of Neurophyiology*, 55, 1328-1339.

[53] Xiao, D. K., Raiguel, S., Marcar, V., & Orban, G. A. (1997). The spatial distribution of the antagonistic surround of MT/V5 neurons. *Cerebral Cortex*, 7, 662-677.

[54] Movshon, J. A., Adelson, E. H., Gizzi, M. S., & Newsome, W. T. (1986). The analysis of moving visual patterns. In C. Chagas, R. Gattass, & C. Gross (Eds.), *Experimental Brain Research Supplementum. II. Pattern recognition mechanisms* (pp. 117-151). New York: Springer-Verlag.

[55] Britten, K. H., Shadlen, M. N., Newsome, W. T. & Movshon, J. A. (1992). The analysis of visual motion: a comparison of neuronal and psychophysical performance. *The Journal of Neuroscience*, 12, 4745-4765.

[56] Gallese, V. (2007). The "conscious" dorsal stream: embodied simulation and its role in space and action conscious awareness. *Psyche*, 13, 1-20.

[57] Siegel, R. M., & Reid, H. L. (1997). Analysis of optic flow in the monkey parietal 7a. *Cerebral Cortex*, 7, 327-346.

[58] Raffi, M., & Siegel, R. M. (2007). A functional architecture of optic flow in the inferior parietal lobule of the behaving monkey. *PLoS ONE*, 2, e200.

[59] Gibson, J.J. (1950). *The perception of the visual world.* Boston MA: Houghton Mifflin.

[60] Warren, W.H., & Hannon, D.J. (1988). Direction of self-motion is perceived from optic flow. *Nature*, 336, 162-163.

[61] Claeys, K. G., Lindsey, D. T., Schutter, E. D., & Orban, G. A. (2003). A higher order motion region in human inferior parietal lobule: evidence from fMRI. *Neuron*, 40, 631-642.

[62] Paradis, A. L., Droulez, J., Cornilleau-Pérès, V., & Poline, J. B. (2008). Processing 3D form and 3D motion: respective contributions of attention-based and stimulus-driven activity. *Neuroimage*, 43, 736-747.

[63] Yamasaki, T., & Tobimatsu, S. (2011). Motion perception in healthy humans and cognitive disorders. In: J. Wu (Ed), *Early detection and rehabilitation technologies for dementia: Neuroscience and biomedical applications*, (pp. 156-161). Hershey, Pennsylvania: IGI Global.

[64] Yamasaki, T., Fujita, T., Kamio, Y., & Tobimatsu, S. (2011). Motion perception in autism spectrum disorder. In: A. M. Columbus (Ed), *Advances in Psychology Research, vol. 82*, (pp. 197-211). New York: Nova Science Publishers.

[65] Yamasaki, T., Muranaka, H., Kaseda, Y., Mimori, Y., & Tobimatsu, S. (2012). Understanding the pathophysiology of Alzheimer's disease and

mild cognitive impairment: A mini review on fMRI and ERP studies. *Neurology Research International,* 2012, 719056.

[66] Galletti, C., Battaglini, P. P, & Fattori, P. (1991). Functional properties of neurons in the anterior bank of the parieto-occipital sulcus of the macaque monkey. *European Journal of Neuroscience*, 3, 452-461.

[67] Galletti, C., Fattori, P., Battaglini, P. P., Shipp, S., & Zeki, S. (1996). Functional demarcation of a border between areas V6 and V6A in the superior parietal gyrus of the macaque monkey. *European Journal of Neuroscience*, 8, 30-52.

[68] Galletti, C., Fattori, P., Gamberini, M., & Kutz, D. F. (1999). The cortical visual area V6: brain locationand visual topography. *European Journal of Neuroscience*, 11, 3922-3936.

[69] Fattori, P., Galletti, C., & Battaglini, P. (1992). Parietal neurons encoding visual space in a head-frame of reference. *Bollettino della Società Italiana di Biologia Sperimentale*, 68, 663-670.

[70] Caminiti, R., Chafee, M. V., Battaglia-Mayer, A., Averbeck, B. B., Crowe, D. A., & Georgopoulos, A. P. (2010). Understanding the parietal lobe syndrome from a neurophysiological and evolutionary perspective. *European Journal of Neuroscience*, 31, 2320-2340.

[71] Braddick, O. J., O'Brien, J. M. D., Wattam-Bell, J., Atkinson, J., & Hartley, T. (2001). Brain areas sensitive to coherent visual motion. *Perception*, 30, 61-72.

[72] Farah, M. J. (1992). Agnosia. *Current Opinion in Neurobiology*, 2, 162-164.

[73] Farah, M. J. (2004). *Visual agnosia.* (2 ed.). Cambridge (MA): The MIT Press.

[74] Catani, M. & Ffytche, D. H. (2005). The rises and falls of disconnection syndromes. *Brain*, 128, 2224-2239.

[75] Damasio, A. R. (1985). Prosopagnosia. *Trends in Neurosciences*, 8, 132-135.

[76] Damasio, A. R., Damasio, H., & Vanhoesen, G. W. (1982). Prosopagnosia-Anatomic Basis and Behavioral Mechanisms. *Neurology*, 32, 331-341.

[77] DeRenzi, E., Perani, D., Carlesimo, G. A., Silveri, M. C., & Fazio, F. (1994). Prosopagnosia can be associated with damage confined to the right hemisphere-an MRI and PET study and a review of the literature. *Neuropsychologia*, 32, 893-902.

[78] Kleinschmidt, A., & Cohen, L. (2006). The neural bases of prosopagnosia and pure alexia: recent insights from functional neuroimaging. *Current Opinion in Neurology*, 19, 386-391.

[79] Zeki, S. (1990). A century of cerebral achromatopsia. *Brain*, 113, 1721-1777.

[80] Meadows, J. C. (1974). Disturbed perception of colours associated with localized cerebral lesions. *Brain*, 97, 615-632.

[81] Damasio, A., Yamada, T., Damasio, H, Corbett, J., & McKee, J. (1980). Central achromatopsia: behavioral, anatomic, and physiologic aspects. *Neurology*, 30, 1064-1071.

[82] Girkin, C. A., & Miller, N. R. (2001). Central disorders of vision in humans. *Survey of Ophthalmology*, 45, 379-405.

[83] Rizzo, M., Nawrot, M., & Zihl, J. (1995). Motion and Shape Perception in Cerebral Akinetopsia. *Brain*, 118, 1105-1127.

[84] Zeki, S. (1991). Cerebral akinetopsia (visual motion blindness). A review. *Brain,* 114, 811-824.

[85] Zihl, J., von Cramon, D., & Mai, N. (1983). Selective disturbance of movement vision after bilateral brain damage. *Brain*, 106, 313-340.

[86] Schenk, T. & Zihl, J. (1997). Visual motion perception after brain damage. 1. Deficits in global motion perception. *Neuropsychologia*, 35, 1289-1297.

[87] Blanke, O., Landis, T., Mermoud, C., Spinelli, L., & Safran, A. B. (2003). Direction-selective motion blinedness after unilateral posterior brain damage. *European Journal of Neuroscience*, 18, 709-722.

[88] Balint, R. (1909). Seelenlähmung des 'Schauens', optische Ataxie, räumliche Störung der Aufmerksamkeit. *Monatsschrift für Psychiatrie und Neurologie*, 25, 51-81.

[89] Jackson, G. M., Shepherd, T., Mueller, S. C., Husain, M., & Jackson, S. R. (2006). Dorsal simultanagnosia: An impairment of visual processing or visual awareness? *Cortex*, 42, 740-749.

[90] Perenin, M. T. & Vighetto, A. (1988). Optic ataxia: a specific disruption in visuomotor mechanisms. I. Different aspects of the deficit in reaching for objects. *Brain*, 111, 643-674.

[91] De Kort, P. (1996). *Neglect, een klinisch onderoek naar halfzijdige verwaarlozing bij patienten met een cerebrale bloeding of infarct.* RijksUniversiteit Groningen.

[92] Bisiach, E. & Vallar, G. (1988). Hemineglect in humans. In: F. Boller & J. Grafman (Eds.), *Handbook of neuropsychology, Vol. 1.* (pp. 195-222). Amsterdam: Elsevier.

[93] Vallar, G. (1993). The anatomical basis of spatial neglect in humans. In: I. H. Robertson & J. C. Marshall (Eds.), *Unilateral Neglect: Clinical and Experimental Studies.* Hove, Sussex: Lawrence Erlbaum Associates.

[94] Robertson, I. H. & Halligan, P. W. (1999). *Spatial neglect: A clinical handbook for diagnosis and treatment.* Hove, East Sussex: Erlbaum.

[95] Blakemore, C., & Campbell, F. W. (1969). On the existence of neurones in the human visual system selectively sensitive to the orientation and size of retinal images. *The Journal of Physiology*, 203, 237-260.

[96] De Valois, K. K., De Valois, R. L., & Yund, E.W. (1979). Responses of striate cortex cells to grating and checkerboard patterns. *The Journal of Physiology*, 291, 483-505.

[97] Holder, G. E., Celesia, G. G., Miyake, Y., Tobimatsu, S., & Weleber, R. G. (2010). International Federation of Clinical Neurophysiology: Recommendations for visual system testing. *Clinical Neurophysiology*, 121, 1393-1409.

[98] Regan, D. (1989). Human brain electrophysiology. *Evoked potentials and evoked magnetic fields in science and medicine.* New York: Elsevier.

[99] Nakamura, A., Kakigi, R., Hoshiyama, M., Koyama, S., Kitamura, Y., & Shimojo, M. (1997). Visual evoked cortical magnetic fields to pattern reversal stimulation. *Brain Research Cognitive Brain Research*, 6, 9-22.

[100] Seki, K., Nakasato, N., Fujita, S., Hatanaka, K., Kawamura, T., Kanno, A., & Yoshimoto, T. (1996). Neuromagnetic evidence that the P100 component of the pattern reversal visual evoked response originates in the bottom of the calcarine fissure. *Electroencephalography and Clinical Neurophysiology*, 100, 436-442.

[101] Shigeto, H., Tobimatsu, S., Yamamoto, T., Kobayashi, T., & Kato, M. (1998). Visual evoked cortical magnetic responses to checkerboard pattern reversal stimulation: a study on the neural generators of N75, P100 and N145. *Journal of Neurological Sciences*, 156, 186-194.

[102] Murray, I. J., Parry, N. R. A., Carden, D., & Kulikowski, J. J. (1987). Human visual evoked potentials to chromatic and achromatic gratings. *Clinical Vision Science*, 1, 231-244.

[103] Porciatti, V., & Sartucci, F. (1999). Normative data for onset VEPs to red–green and blue–yellow chromatic contrast. *Clinical Neurophysiology*, 110, 772-781.

[104] Tobimatsu, S., & Kato, M. (1998). Multimodality visual evoked potentials in evaluating visual dysfunction in optic neuritis. *Neurology*, 50, 715-718.

[105] Tobimatsu, S., Tomoda, H., & Kato, M. (1995). Parvocellular and magnocellular contributions to visual evoked potentials in humans: stimulation with chromatic and achromatic gratings and apparent motion. *Journal of Neurological Sciences*, 134, 73-82.

[106] Fujita, T., Yamasaki, T., Kamio, Y., Hirose, S., & Tobimatsu, S. (2011). Parvocellular pathway impairment in autism spectrum disorder: Evidence from visual evoked potentials. *Research in Autism Spectrum Disorders*, 5, 277-285.

[107] Gutschalk, A., Patterson, R. D., Rupp, A., Uppenkamp, S., & Scherg, M. (2002). Sustained magnetic fields reveal separate sites for sound level and temporal regularity in human auditory cortex. *Neuroimage*, 15, 207-216.

[108] Yamasaki, T., Taniwaki, T., Tobimatsu, S., Arakawa, K., Kuba, H., Maeda, Y., Kuwabara, Y., Shida, K., Ohyagi, Y., Yamada, T., & Kira, J. (2004). Electrophysiological correlates of associative visual agnosia lesioned in the ventral pathway. *Journal of Neurological Sciences*, 221, 53-60.

[109] Bentin, S., Allison, T., Puce, A., Perez, A., & McCarthy, G. (1996). Electrophysiological studies of face perception in humans. *Journal of Cognitive Neuroscience*, 8, 551-565.

[110] Bötzel, K., Schulze, S., & Stodieck, R.G. (1995). Scalp topography and analysis of intracranial sources of face-evoked potentials. *Experimental Brain Research*, 104, 135-143.

[111] Eimer, M. (2000). Effects of face inversion on the structural encoding and recognition of faces-evidence from event-related brain potentials. *Cognitive Brain Research*, 10, 145-158.

[112] Eimer, M. (2000). Attentional modulations of event-related brain potentials sensitive to faces. *Cognitive Neuropsychology*, 17, 103-116.

[113] Halgren, E., Raij, T., Marinkovic, K., Jousmaki, V., & Hari, R. (2000). Cognitive response profile of the human fusiform face area as determined by MEG. *Cerebral Cortex*, 10, 69-81.

[114] Liu, J., Higuchi, M., Marantz, A., & Kanwisher, N. (2000). The selectivity of the occipitotemporal M170 for faces. *Neuroreport*, 11, 337-341.

[115] Rossion, B., Gauthier, I., Tarr, M. J., Despland, P. A., Bruyer, R., Linotte, S., & Crommelinck, M. (2000). The N170 occipito-temporal component is enhanced and delayed to inverted faces but not to inverted objects: an electrophysiological account of face-specific processes in the human brain. *Neuroreport*, 11, 69-74.

[116] Schendan, H. E., Ganis, G., & Kutas, M. (1998). Neurophysiological evidence for visual perceptual categorization of words and faces within 150 ms. *Psychophysiology*, 35, 240-251.

[117] Caldara, R., Rossion, B., Bovet, P., & Hauert, C. A. (2004). Event-related potentials and time course of the ''other-race'' face classification advantage. *Neuroreport*, 15, 905-910.

[118] Rossion, B., Joyce, C. A., Cottrell, G. W., & Tarr, M. J. (2003). Early lateralization and orientation tuning for face, word, and object processing in the visual cortex. *Neuroimage*, 20, 1609-1624.

[119] de Haan, M., Pascalis, O., & Johnson, M. H. (2002). Specialization of neural mechanisms underlying face recognition in human infants. *Journal of Cognitive Neuroscience*, 14, 199-209.

[120] Schweinberger, S. R., & Sommer, W. (1991). Contributions of stimulus encoding and memory search to right hemisphere superiority in face recognition: Behavioural and electrophysiological evidence. *Neuropsychologia*, 29, 389-413.

[121] Simon, G. D. A., Petit, L., Bernard, C., & Rebai, M. (2007). N170 ERPs could represent a logographic processing strategy in visual word recognition. *Behavioral and Brain Functions*, 3, 21.

[122] Maurer, U., Brandeis, D., & McCandliss, B. (2005). Fast, visual specialization for reading in English revealed by the topography of the N170 ERP response. *Behavioral and Brain Functions*, 1, 13.

[123] Caldara, R., Jermann, F., Lopez Arango, G., & Van der Linden, M. (2004). Is the N400 category-specific? A face and language processing study. *Neuroreport*, 15, 2589-2593.

[124] Bentin, S., Mouchetant-Rostaing, Y., Giard, M. H., Echallier, J. F., & Pernier, J. (1999). ERP manifestations of processing printed words at different psycholinguistic levels: Time course and scalp distribution. *Journal of Cognitive Neuroscience*, 11, 235-260.

[125] Gerschlager, W., Lalouschek, W., Lehrner, J., Baumgartner, C., Lindinger, G., & Lang, W. (1998). Language-related hemispheric asymmetry in healthy subjects and patients with temporal lobe epilepsy as studied by event-related brain potentials and intracarotid amobarbital test. *Electroencephalography and Clinical Neurophysiology*, 108, 274-282.

[126] Arakawa, K., Tobimatsu, S., Kato, M., & Kira, J. (1999). Parvocellular and magnocellular visual processing in spinocerebellar degeneration and Parkinson's disease: an event-related potential study. *Clinical Neurophysiology*, 110, 1048-1057.

[127] Newsome, W. T., & Paré, E. B. (1988). A selective impairment of motion perception following lesions of the middle temporal area (MT). *The Journal of Neuroscience*, 8, 2201-2211.

[128] Niedeggen, M., & Wist, E. R. (1999). Characteristics of visual evoked potentials generated by motion coherence onset. *Brain Research Cognitive Brain Research*, 8, 95-105.

[129] Morrone, M. C., Tosetti, M., Montanaro, D., Fiorentini, A., Cioni, G., & Burr, D. C. (2000). A cortical area that responds specifically to optic flow, revealed by fMRI. *Nature Neuroscience*, 3, 1322-1328.

[130] Tsai, P. H., & Mendez, M. F. (2009). Akinetopsia in the posterior cortical variant of Alzheimer disease. *Neurology*, 73, 731-732.

[131] Mendez, M. F., Mendez, M. A., Martin, R., Smyth, K. A., & Whitehouse, P. J. (1990). Complex visual disturbances in Alzheimer's disease. *Neurology*, 40, 439-443.

[132] Cronin-Golomb, A., Corkin, S., Rizzo, J. F., Cohen, J., Growdon, J. H., & Banks, K. S. (1991). Visual dysfunction in Alzheimer's disease relation to normal aging. *Annals of Neurology*, 29, 41-52.

[133] Butter, C. M., Trobe, J. D., Foster, N. L., & Brent, S. (1996). Visual-spatial deficits explain visual symptoms in Alzheimer's disease. *American Journal of Ophthalmology*, 122, 97-105.

[134] Mapstone, M., Steffenella, T. M., & Duffy, C. J. (2003). A visuospatial variant of mild cognitive impairment: getting lost between aging and AD. *Neurology*, 60, 802-808.

[135] Yamasaki, T., Horie, S., Himeno, E., Nakamura, N., Ohyagi, Y., Kira, J. I., & Tobimatsu, S. (2011). A deficit of dorsal stream function in patients with amnestic mild cognitive impairment. *Clinical Neurophysiology*, 122, 20.

[136] Tobimatsu, S., Goto, Y., Yamasaki, T., Tsurusawa, R., & Taniwaki, T. (2004). Non-invasive evaluation of face and motion perception in humans. *Journal of Physiological Anthropology and Applied Human Science*, 23, 273-276.

[137] Tobimatsu, S., Goto, Y., Yamasaki, T., Tsurusawa, R., & Taniwaki, T. (2006). An integrated approach to face and motion perception in humans. *Clinical Neurophysiology*, S59, 43-48.

[138] Tobimatsu, S., Goto, Y., Yamasaki, T., Nakashima, T., Tomoda, Y., & Mitsudome, A. (2008). Visual ERPs and cortical function. In: A. Ikeda, & Y. Inoue (Eds.), *Progress in Epileptic disorders, Volume 5, Event-related potentials in patients with epilepsy: from current state to future prospects*, (pp. 37-48). Paris: John Libbey Eurotext.

[139] Frith, U., & Happé, F. (2005). Autism spectrum disorder. *Current Biology*, 15, 786-790.

[140] Happé, F. G. (1996). Studying weak central coherence at low levels: children with autism do not succumb to visual illusions: A research note. *Journal of Child Psychology and Psychiatry*, 37, 873-877.

[141] Happé, F., & Frith, U. (2006). The weak coherence account: detail-focused cognitive style in autism spectrum disorders. *Journal of Autism and Developmental Disorders*, 36, 5-25.

[142] Jolliffe, T., & Baron-Cohen, S. (1997). Are people with autism and Asperger syndrome faster than normal on the Embedded Figures Test? *Journal of Child Psychology and Psychiatry*, 38, 527-534.

[143] Spencer, J., O'Brien, J., Riggs, K., Braddick, O., Atkinson, J., & Wattam-Bell, J. (2000). Motion processing in autism: evidence for a dorsal stream deficiency. *Neuroreport*, 11, 2765-2767.

[144] Milne, E., Swettenham, J., Hansen, P., Campbell, R., Jeffries, H., & Plaisted, K. (2002). High motion coherence thresholds in children with autism. *Journal of Child Psychology and Psychiatry*, 43, 255-263.

[145] Bertone, A., Mottron, L., Jelenic, P., Faubert, J. (2003). Motion perception in autism: a "complex" issue. *Journal of Cognitive Neuroscience*, 15, 218-225.

[146] Dakin, S., & Frith, U. (2005). Vagaries of visual perception in autism. *Neuron*, 48, 497-507.

[147] Mottron, L., & Burack, J.A. (2001). Enhanced perceptual functioning in the development of autism. In: J. A. Burack, T. Charman, N. Yirmiya, & P. R. Zelazo. (Eds.), *The Development of Autism: Perspectives from Theory and Research* (pp. 131-148). Mahwah: Lawrence Erlbaum.

[148] Bertone, A., & Faubert, J. (2006). Demonstrations of decreased sensitivity to complex motion information not enough to propose an autism-specific neural etiology. *Journal of Autism and Developmental Disorders*, 36, 55-64.

[149] Koh, H. C., Milne, E., & Dobkins, K. (2010). Contrast sensitivity for motion detection and direction discrimination in adolescents with autism spectrum disorders and their siblings. *Neuropsychologia*, 48, 4046-4056.

[150] Yamasaki, T., Fujita, T., Ogata, K., Goto, Y., Munetsuna, S., Kamio, Y., & Tobimatsu, S. (2011). Electrophysiological evidence for selective impairment of optic flow perception in autism spectrum disorder. *Research in Autism Spectrum Disorders*, 5, 400-407.

In: Visual Cortex: Anatomy, Functions … ISBN: 978-1-62100-948-1
Editors: J.M. Harris et al. pp. 69-98 © 2012 Nova Science Publishers, Inc.

Chapter 3

PLASTICITY OF VISUAL CORTICAL CIRCUITRIES IN ADULTHOOD

José Fernando Maya-Vetencourt
Scuola Normale Superiore, Neurobiology Laboratory
Neuroscience Institute, Via Moruzzi 1, CNR, Pisa 56100 Italy.

ABSTRACT

The formation of neural circuitries in the brain relies on a tight interaction between genes and environment. As intrinsic factors mediate the initial assembly of synaptic circuitries, the nervous system begins to process sensory information thus creating neuronal representations of the external world that are continuously modified by experience. Sensory experience, in fact, modifies the structural and functional architecture of the nervous system in response to changing environmental conditions throughout life. This phenomenon is particularly evident during early stages of development when experience drives the consolidation of synaptic connectivity but the reorganization of neural circuitries continues in adulthood, as for instance, in response to learning, loss of sensory input or trauma. How does experience modify synaptic circuitries in the visual system? This will be one of the topics I shall concentrate on inthis chapter.

The notion that neural circuitries in the adult brain can change in response to experience has become a major conceptual subject in modern neuroscience.Althoughearly studies dealt with plasticity of the developing nervous system,the study of adult brain plasticity and the identification of molecular and cellular mechanisms at the basis of these plastic phenomena are current challenges in the neuroscience field. This chapter outlines physiological processes that underlie neuronal plasticity in the visual cortex and provides an in-depth discussion about the enhancement of plasticity as a strategy for brain repair in adulthood. Given that experience-dependent changes of brain functions depend, at least partially, on the expression of genes that have evolved to meet specific environmental demands, molecular mechanisms that lie behind processes of neuronal plasticity shall also be addressed. How intracellular signal transduction pathways associated to sensory experience regulate epigenetic modifications of chromatin structure that underlie visual cortical plasticity is also a novel concept that will be considered.

INTRODUCTION

Neural Circuitries Formation Relies on a Tight Interaction between Genes and Environment

The nervous system is highly sensitive to experience during early stages of development (critical period) but this phase of heightened plasticity decreases with age. Although intrinsic factors mediate the initial assembly of neuronal circuitries in the brain, patterns of neural connectivity are shaped by experience during early postnatal life. Spontaneous electrical activity and activity associated to sensory experience drive the formation of neural circuitries, promoting for instance, the establishment, elimination and rearrangements of synaptic connections. In the visual system, intrinsic signals guide the projection of retinal inputs to different subcortical and cortical structures, creating the neuronal substrate for a complete retinotopic map of the visual field. Early electrophysiological studies revealed that the segregation of retinal projections into eye specific patches at the level of the lateral geniculate nucleus occurs during embryonic life (Shatz and Stryker, 1988; Snider et al., 1999). This anatomical segregation is initially driven by spontaneously generated synchronous firing of action potentials in the mammalian retina before eye opening (Katz and Shatz, 1996). Such spontaneous patterns of electrical activityhave, in fact, been reported in several species: rats (Galli and Maffei, 1988), ferrets (Meister et al, 1991) and mice (Bansal et al., 2000). Other functional characteristics in the visual system such as rough orientation selectivity of cortical neurons are also present before visual experience (Wiesel and Hubel, 1974), indicating that the initial anatomical and functional organization of the visual cortex relies, at least partially, on innate processes. However, once basic patterns of neural circuitries are formed, sensory experience plays a key role in the refinement of synaptic connectivity in the brain. These findings highlight the fact that neuronal plasticity is particularly high during early stages of development (critical period for visual cortex plasticity) but that it decreases late in life (Berardi et al., 2000) when major alterations of neural circuitries structures no longer take place in response to experience.

Visual Cortical Plasticity: A Classic Neurobiological Model

Once basic patterns of neuronal circuitries are formed, an experience-dependent reorganization of eye-specific inputs is themajor mechanism by which neuronal connectivity is established in the developing visual system. Using the monocular deprivation paradigm, pioneering electrophysiological studies in cats and monkeys clearly demonstrated that the ocular dominance distribution in the binocular area of the primary visual cortex dramatically changes in response to unilateral eyelid suture during the critical period. Monocular deprivation during early life markedly modifies visual cortex responsiveness, as it causes a reduction in the number of cells driven by the deprived eye that parallels an increment in the number of neurons activated by the open eye. In addition, the deprived eye becomes amblyopic: its visual acuity (spatial resolution) and contrast sensitivity are markedly impaired (Wiesel and Hubel, 1963a; LeVay et al, 1980). Furthermore, eyelid suture leads to a reduced arborisation of geniculocortical terminals serving the deprived eye together with an increased spread of terminals serving the open eye (Wiesel and Hubel, 1963b). Because sensory deprivation does not cause amblyopia in adulthood these findings illustrate a typical example of a critical period for visual cortex plasticity. The importance of sensory experience in development of the human brain is epitomized by cases of strabismic or anisometric children who received no clinical treatment during early development. These two pathological conditions cause a marked impairment of normal visual functions (amblyopia) that is irreversible in adult life (Holmes and Clarke, 2006). The recovery of function after long-term sensory deprivation has long been a subject of attention with the prospect of finding therapies for human amblyopia. Therefore, strategies that restore plasticity in the adult visual system offer particular promise for possible future applications at clinical level.

Another classical paradigm used to assess the impact of sensory experience in the functional maturation of the visual system is dark rearing (i.e., rearing animals in total darkness from birth). The absence of visual experience during early life delays the maturation of the striate cortex. In fact, the visual system of dark-reared animals displays reduced levels of spatial acuity and longer latencies of response to visual stimuli (Fagiolini et al., 1994) as compared to animals reared under normal conditions. Furthermore, visual cortical neurons have larger receptive field sizes and show alterations in the size and density of dendritic spines, which are shorter and fewer in dark reared animals with respect to normally reared counterparts (Wallace and Bear, 2004). It is also worthnoting that dark rearing also extents the critical period and prolongs plasticity far beyond its normal limits.

Modifications of Neural Circuitries in Adult Life

The notion that cortical circuitries in the adult brain can be modified in response to experience has become a major conceptual subject in modern neuroscience. Although early studies dealt with plasticity in the developing brain, the study of adult cortical plasticity and the identification of molecular mechanisms that lie behind these plastic phenomenaare hot spots in this field of research. How intracellular signal transduction pathways associated to experience regulate changes of chromatin structure that underlie phenomena of plasticity in the adult brain is a concept of particular interest on the field. Epigenetic mechanisms that exert a long-lasting control of gene expression by altering chromatin structure rather than changing the DNA sequence itself are, in fact, a common theme in long-term plastic changes from vertebrates to invertebrates. The core histones, for instance, are subject to a variety of post-translational modifications such as acetylation, methylation, phosphorylation and ubiquitination, all of which have recently emerged as a conserved process by which the nervous system accomplishes the induction of plasticity in adult life.

PHYSIOLOGICAL MECHANISMS UNDERLYING PLASTICITYIN THE VISUAL SYSTEM

Experimental research over the last decades has shed light onto the mechanisms that regulate experience-dependent plasticity in the mammalian visual cortex. A number of different signaling molecules, including neurotrophic factors, extracellular matrix molecules, myelination, N-methyl-D-aspartate (NMDA)-type glutamate signaling, γ-aminobutyric acid (GABA)-mediated inhibitory transmission and ascending projection systems, have been recognized as important regulators of visual cortical plasticity.

Inhibitory Processes Regulate the Time-Course of the Critical Period in the Visual System

Intracortical inhibitory processes play a key role in the regulation of visual cortex plasticity (Hensch, 2005). The maturation of GABA-mediated inhibition sets the threshold for the start of the critical period for experience-dependent plasticity. In fact, transgenic mice with reduced levels of intracortical inhibition, due to the absence of one isoform of the GABA synthesizing enzyme (glutamate decarboxylase of 65 kDa - GAD65), show no susceptibility to monocular deprivation during their entire lifespan (Hensch et al., 1998; Fagiolini and Hensch, 2000). Most importantly, the impairment of experience-dependent plasticity in these animals is rescued by enhancing inhibitory transmission by means of the benzodiazepine Diazepam (agonist of GABAA receptors) (Fagiolini et al., 2004). Thus, a reduction of inhibitory transmission during early postnatal life halts the onset of critical period, an effect that is rescued by enhancing inhibition. A second inhibitory threshold, which causes the closure of the critical period, is reached over postnatal development as well. This was demonstrated in transgenic animals over-expressing BDNF in forebrain regions, in which an accelerated maturation of intracortical inhibitory circuitries determines a precocious termination of the critical period for ocular dominance plasticity(Huang et al., 1999). In summary, an initial threshold of inhibition triggers a critical period in which neural networks in the visual system are highly susceptible to sensory experience, and a second inhibitory threshold signals the end of this phase of heightened plasticity.

In line with this notion, it is possible to restore plasticity late in life by reducing levels of inhibition. Studies performed in the laboratory of LambertoMaffei revealed that a pharmacological reduction of inhibitory transmission by infusion of mercaptopropionic acid (MPA, an inhibitor of GABA synthesis) or Picrotoxin (a GABAA receptor antagonist) into the rat visual cortex effectively reactivates ocular dominance plasticity in adulthood (Harauzov et al., 2010).

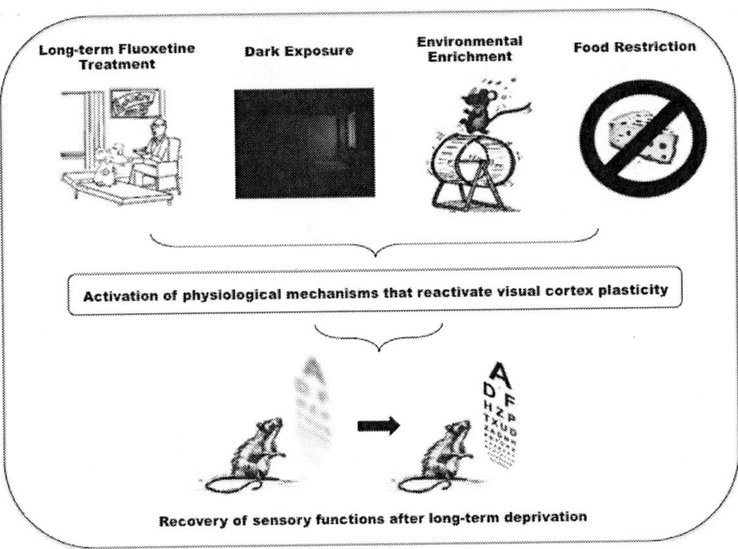

Figure 1. Environmental enrichment, long-term treatment with the antidepressant fluoxetine, visual deprivation by dark exposure and food restriction, all promote the recovery of normal visual function in adult amblyopic rats by altering the balance of inhibition and excitation in the visual system. In particular, dark exposure decreases the expression of GABAA receptors in parallel to a shift of NMDA receptors subunit composition (He et al., 2006). Instead, environmental enrichment (Sale et al., 2007; Baroncelli et al., 2010), antidepressant treatment (Maya-Vetencourt et al., 2008; 2011) and food restriction (Spolidoro et al., 2011) consistently reduce extracellular levels of GABA without affecting those of glutamate. Moreover, these experimental strategies promote BDNF expression while enhancing LTP occurrence. This is in line with the observation that a reduction of intracortical inhibitory transmission, a pharmacological strategy that reinstates ocular dominance plasticity in adulthood (Harauzov et al., 2010), not only causes the removal of extracellular matrix components that are inhibitory for plasticity but also enhances BDNF levels. The pharmacological removal of extracellular matrix components that are inhibitory for plasticity reinstates susceptibility to monocular deprivation in adulthood and promotes the recovery of visual functions in adult amblyopic animals (Pizzorusso et al., 2002; 2006).

This is in agreement with the fact that experimental paradigms such as: dark exposure (He et al., 2006), environmental enrichment (Sale et al., 2007; Baroncelli et al., 2010), chronic treatment with the antidepressant fluoxetine (Maya-Vetencourt et al., 2008; 2011; Chen et al., 2011) and food restriction (Spolidoro et al., 2011), all promote plasticity in adult life by reducing the inhibitory/excitatory ratio in the visual cortex (see Figure 1). How does a shift of the inhibitory/excitatory balance trigger plasticity in the adult visual system? Although not yet clear, a reduction of inhibition as compared to excitation is likely to decrease the threshold for visual cortical neurons to be driven by electrical activity thus increasing the probability for visual inputs to drive experience-dependent modifications of synaptic transmission in adulthood.

Experience-Dependent Retinogeniculocortical Transfer of Otx2

A newfactor recently found to control visual cortex plasticity during development is the retinogeniculocortical transfer of the homeoptrotein Otx2. Recent experimental data obtained in the laboratories of Alain Prochiantz and Takao Hensch revealed that Otx2 is transported from the retina to the visual cortex and regulates critical period plasticity by promoting the maturation of parvalbumin-positive GABAergic interneurons. The authors found that intracortical delivery of the recombinant Otx2 protein in mice before the onset of the critical period prematurely shifts visual cortex responsiveness towards the open eye after deprivation by eyelid suture. This suggests that Otx2 triggers the critical period onset by enhancing levels of inhibition (Sugiyama et al., 2008), as weak inhibition at early stages of development prevents the occurrence of experience-dependent plasticity in the visual cortex. In support ofthis notion, excessive spike firing of cortical neurons was effectively reduced by exogenous administration of Otx2 in the mouse visual cortex before the onset of the critical period (Sugiyama et al., 2008), just like benzodiazepines treatment reduces this immature property in the visual system of wild-type animals (Hensch et al., 1998).

That Otx2 is required for the maturation of parvalbumin-positive interneurons and the occurrence of critical period plasticity, was confirmed in conditional Otx2 knockout animals. In fact, a reduction of inhibitory transmission is observed in slices of Otx2 knockout mice. Moreover, ocular dominance plasticity is lessened in these transgenic animals, as expected from the impairment of inhibition; and this effect is rescued by enhancing GABAA receptor currents by benzodiazepines treatment (Sugiyama et al., 2008), as similarly observed in GAD65 knockout mice(Hensch et al., 1998). Taken together, these observations suggest that the retinogeniculocortical transfer of a molecular messenger is a mechanism by which sensory experience signals the time-course of the critical period. These findings also have important implications forprocesses of adult visual cortex plasticity. Normal visual functions gradually degrade upon aging and this phenomenon is accompanied by a deterioration of intracortical inhibitory processes (Leventhal et al., 2003; Betts et al., 2005). Therefore, the use of Otx2 as a signaling molecule that is likely to compensate aging-related deficits of inhibitory transmission seems to be a potential strategy to promote the recovery of sensory functions during senescence.

Plasticity Involves the Remodeling of Extracellular Matrix Components

There is evidence that removal of particular components of the extracellular environment in the central nervous system is necessary for experience-dependent plasticity to occur. The proteolytic activity of tissue plasminogen activator (tPA), for instance, increases after monocular deprivation during the critical period (Mataga et al., 2002) and plays a permissive role for visual cortex plasticity as indicated by the fact that a pharmacological inhibition of the tPA attenuates shifts of ocular dominance following deprivation by eyelid suture in mice(Mataga et al., 1996). Intracortical blockade of tPA also prevents recovery of cortical function during reverse occlusion in kittens (Müller and Griesinger, 1998). Furthermore, experience-dependent plasticity during the critical period is impaired in tPA-knockout mice and such impairment is rescued by exogenous tPA administration. This is consistent with a key role for tPA in structural plasticity of the developing visual system in mice (Mataga et al., 2004). For instance, changes of binocularity in response to monocular deprivation are accompanied by rapid structural changes at the level of dendritic spines in deep and superficial layers of the primary visual cortex (Oray et al., 2004), an effect that is mimicked by exogenous tPA administration.

Additional evidence confirming the inhibitory action of the extracellular environment in visual cortical plasticity came from the work of TommasoPizzorusso, LambertoMaffei and James Fawcett, who found that condroitinsulphate proteoglycans (CSPGs: glycoproteins that are major components of the extracellular matrix) are inhibitory for experience-dependent plasticity in the rat. The degradation of sugar chains of CSPGs by exogenous administration of the enzyme chondroitinase-ABC, indeed, reactivates ocular dominance plasticity in the adult(Pizzorusso et al., 2002). This was an exciting finding as it suggested a potential pharmacological treatment for the recovery of sensory functions in adulthood, which was demonstrated by the fact that chondroitinase-ABC treatment promotes structural and functional recovery from visual deficits in long-term deprived rats (Pizzorusso et al., 2006). Although the effectiveness of chondroitinase-ABC treatment has been consistently observed in rodents, its action is not as clear in the cat visual system, where no functional recovery of visual functions has been detected (Frank Sengpiel, personal communication). It is important to remark that degradation of CSPGs may alter the ratio of inhibitory/excitatory transmission in the visual cortex, as these glycoproteins condense in perineuronal nets (PNNs) mainly around inhibitory interneurons. Whether or not chondroitinase-ABC treatment has an impact on the inhibitory tone in the visual system is currently unknown. Another interesting notion is that removing CSPGs may provide a permissive environment for structural plasticity (e.g., by modifying dendritic spine dynamics) and associated functional modifications in the visual cortex. In keeping with this, an increased dynamics of dendritic spines correlates with NMDA-receptor dependent functional changes during experience-dependent plasticity in the rodent visual system (Tropea et al., 2010).

The Maturation of Intracortical Myelination Down-Regulates Plasticity

A critical role for myelin-associated processesin visual cortical plasticity has also been recognized. In particular, proteins that compose the myelin sheath such as: (i) Nogo-A and (ii) myelin-associated glycoprotein (MAG), control plasticity via their binding to the receptor NgR. A series of refinedexperiments conducted in the laboratories of Stephen Strittmatter and Nigel Daw provided insights into how NgR signaling contributes to the time-course of critical period plasticity. The maturation of intracortical myelination was found to correlate with the end of the critical period and ocular dominance plasticity persisted well into adulthoodin mice knockout for NgR (McGee et al., 2005). Moreover, transgenic animals lacking Nogo-A displayed a similar susceptibility to monocular deprivation thus confirming that NgR-dependent mechanisms participate directly in restricting experience-dependent plasticity of the visual system. More recently, studies performed in the laboratories of Carla Shatz and Marc Tessier-Lavigne have clearly demonstrated that the paired immunoglobulin-like receptor B (PirB) shows high affinity for Nogo-A and MAG, the signaling of which is inhibitory for axonal regeneration (Atwal et al., 2008). In keeping with this, PirB restricts ocular dominance plasticity in the developing visual cortex (Syken et al., 2006).

Neurotrophic Factors Drive Experience-Dependent Forms of Plasticity

The role of neurotrophins in the context of plasticity was initially explored in the early 1990s by pioneering studies in the laboratory of LambertoMaffei, whose initial attention was directed towards NGF. The authors demonstrated that exogenous NGF administration prevents the loss of visual acuity and the shift of ocular dominance caused by monocular deprivation in the rat visual system (Domenici et al., 1991; Maffei et al., 1992). This observation led them to propose the hypothesis that ocular dominance plasticity involves the competition between geniculocortical projections from either eye for a neurotrophic factor, which is produced and released by visual cortical neurons in an activity-dependent manner.

Several lines of research later confirmed this hypothesis. The NGF-induced regulation of ocular dominance plasticity depends on afferent electrical activity (Caleo et al., 1999) and the inactivation of NGF signaling by specific antibodies at the level of the cortex not only impairs the anatomical and functional development of the visual system (Berardi et al., 1994), but also prolongs the critical period beyond its normal limits (Domenici et al., 1994). In addition, NGF administration and activation of trkA receptors prevent the shift of visual cortex responsiveness after eyelid suture during the critical period (Pizzorusso et al., 1999). Although the effectiveness of NGF to prevent the effects of monocular deprivation has been consistently observed in rodents, its action is not as clear in the cat visual system. Intraventricular, but not intracortical, infusion of NGF attenuates the effects of eyelid suture in kittens (Carmignoto et al., 1993; Galuske et al., 1996), indicating that NGF does not act directly on visual cortical neurons, but affects other systems. The intraventricular NGF delivery is likely to activate subcortical structures bearing NGF receptors, such as afferents from the cholinergic system in the basal forebrain. The primary NGF receptor trkA, indeed, is mostly, but not exclusively, expressed in cholinergic fibres.

A role for BDNF as a mediator of plasticity in the nervous system was hypothesized upon the observation that its expression is activity-dependent and widespread in the brain. The fact that BDNF infusion desegregates ocular dominance columns in the kitten visual system (Cabelli et al., 1995), suggested that a competition for limiting amounts of BDNF might regulate the functional layout of geniculocortical projections. A similar effect was evidenced by infusion of NT-4, which also binds to the trkB receptor. In line with this notion, BDNF infusion in the rat visual cortex blocks the physiological effects of monocular deprivation during the critical period (Lodovichi et al., 2000), this phenomenon being observed in the kitten visual system as well (Galuske et al., 1996). Furthermore, depletion of trkB ligands by exogenous trkB-IgG interferes with the anatomical segregation of ocular dominance columns in kittens (Cabelli et al., 1997).

Striking evidence for thc role of this neurotrophin in regulating the critical period came from studies in transgenic mice over-expressing BDNF. A precocious BDNF expression accelerates the development of visual acuity and the time course of experience-dependent plasticity, these events being mediated by an accelerated maturation of intracortical inhibition (Huang et al., 1999). More recently, a key role for BDNF-trkB signaling in restoring ocular dominance plasticity in adulthood has been reported. Intracortical infusion of the neurotrophin in the rat visual cortexwas actually found to reinstate susceptibility to monocular deprivation in adult life (Maya-Vetencourt et al., 2008) whereas the impairment of BDNF-trkB signaling blocks the reinstatement of plasticity caused by the antidepressant fluoxetine in the adult visual system (Maya-Vetencourt et al., 2011).

How does BDNF promote functional modifications in the nervous system? Studies performed on cats and rodents during the critical period, indicate that electrical activity and neurotrophin signaling cooperate to set in motion different protein kinases: cAMP-dependent protein kinase (PKA) (Beaver et al., 2001), extracellular-signal-regulated kinase (ERK1/2) (Di Cristo et al., 2001) and Ca^{2+}/calmodulin-dependent protein kinase II (CaMKII) (Taha et al., 2002). The activation of such intracellular pathways stimulates the up-regulation of specific transcription factors which, in turn, promote gene expression. A very well-known activity-dependent mechanism is the activation of the transcription factor CREB, which triggers the expression of genes under control of the cAMP-response element (CRE) promoter, thus allowing plasticity to occur (Pham et al., 1999; Mower et al., 2002). The recent observation that a gradual increase of BDNF expression following cocaine withdrawal facilitates long-term potentiation of synaptic transmission in pyramidal neurons by suppressing GABAergic transmission in the rat prefrontal cortex (Lu et al., 2010), suggests that a shift of the intracortical inhibitory/excitatory balance may, at least, be one of the mechanisms by which the neurotrophin could also promote phenomena of plasticity in the adult nervous system.

Neurotransmission Mediated by NMDA-Type Glutamate Receptors (NMDARs)

A role for the switch of NMDARs subunits composition in driving the closure of the critical period was initially suggested upon the observation that it is regulated by age and experience. This phenomenon parallels the functional development of the visual system in cats and rodents (Chen et al., 2000; Quinlan et al., 1999) and correlates with a change in the kinetics of NMDARs mediated currents (Carmignoto et al., 1992; Flint et al., 1997). NMDARs subunits composition varies from an increased expression of receptors containing the NR2B subunit to a progressive inclusion of the subunit NR2A over development, this molecular event being paralleled by a progressive shortening of NMDARs currents (Carmignoto et al., 1992; Flint et al., 1997).

Suchexperimental observations raised the notion that a lack of the NR2A subunit could prolong the critical period beyond its normal limits. However, a normal sensitivity to deprivation by eyelid suturewas found in transgenic mice with deletion of the NR2A subunit. Moreover, the effects of monocular deprivation in these animals were restricted to the typical critical period (Fagiolini et al., 2003). Therefore, the late onset and experience-dependent NR2A expression pattern seems not to be causally linked to the termination of the critical period in the visual cortex (Hensch et al., 2005). Instead, this molecular event appears to be a correlate of experience-dependent plastic phenomena in the visual system.

The observation that blockade of NMDARs inhibits the effects of monocular deprivation in kittens supports the idea that experience-dependent NR2A expression seems to play a permissive role for modifications of synaptic transmission(Bear et al., 1990; Roberts et al., 1998). Extensive research carried out in the laboratory of Mark Bear has providedinsights into the neural mechanisms that mediate the alterations of visual cortex responsiveness following monocular deprivation. Modifications of binocularity induced by eyelid suture during the critical period in mice are mediated by the combination of two different processes: an initial input-specific weakening of synapses from the deprived eye and a later strengthening of synapses from the open eye (Frenkel and Bear, 2004). A reduction of the NR2A/2B ratio has been suggested to modulate these mechanisms of experience-dependent plasticity. The rapid depression of inputs from the deprived eye that is normally observed in juvenile mice after monocular deprivation is absent in transgenic animals lacking the NR2A subunit. Instead, these mice exhibit a precocious potentiation of open eye inputs(Cho et al., 2009). These findings support a permissive role for modifications of NMDARs subunit composition in the compensatory potentiation of non-deprived inputs.

Modifications of Synaptic Transmission in the Visual Cortex

Changes of synaptic transmission in response to sensory experience have long been studied in terms of long-term potentiation (LTP) and long-term depression (LTD). What role do LTP and LTD play in ocular dominance plasticity? A theoretical framework for this is provided by the Bienenstock-Cooper-Munro (BCM) theory, which was developed to explain features of experience-dependent plasticity in the kitten visual cortex and describes the extent of LTP and LTD occurrence as a function of spiking activity in postsynaptic cells (Bienenstock et al., 1982). It predicts that under conditions of low postsynaptic activity, active afferents are depressed, whereas under high postsynaptic firing, those afferents are strengthened.

In this context, the initial weakening of visual inputs from the deprived eye that occurs after brief monocular deprivation may be mediated by homosynaptic mechanisms of LTD. This was demonstrated by using an approach in which *in vivo* occurring LTD occludes LTD induction *ex vivo*. LTD occurrence in visual cortical slices of young animals is, indeed, saturated after eyelid suture for 24 hours (Heynen et al., 2003).

The weak activation of NMDARs mediated by a decreased afferent activity from the deprived-eye causes a low-amplitude rise in Ca^{2+}concentration that activates postsynaptic protein phosphatases. These proteins drive dephosphorylation and internalization of AMPA receptors

(AMPARs) (Heynen et al., 2003), thereby decreasing sensitivity to glutamate, which causes a long-lasting depression of synaptic transmission from the deprived-eye.

The BCM theory also predicts that the late open-eye strength potentiation following monocular deprivation is mediated by a homosynaptic mechanism of LTP. Although specific mechanisms by which open eye potentiation occurs remain unclear, correlated synaptic inputs from the not-deprived eye are likely to activate NMDARs, which mediate high amplitude rises in Ca^{2+}levels. Ca^{2+}-activated protein kinases eventually increase AMPARs insertion at postsynaptic level, thus enhancing sensitivity to glutamate, which allows a long-lasting potentiation of synaptic transmission. This notion is supported by the occurrence of NMDAR-dependent forms of LTP in the rat visual system (Kirkwood et al., 1993) and by the fact that ocular dominance plasticity is impaired in transgenic mice lacking the protein kinase CaMKII, a key player in LTP expression (Gordon et al., 1996; Taha et al., 2002).

Metaplasticity: Mechanisms That Regulate Synaptic Plasticity

Although the BCM theory provides an elegant model that explains how deprived-eye depression and open-eye potentiation following eyelid suture could be achieved in terms of homosynaptic LTD and LTP, it does not account for the difference in the temporal dynamics of these processes.

The increase of the open-eye strength occurs after 5 days of monocular deprivation, while depression of deprived-eye inputs occurs within the first 3 days of it. How are we to understand this phenomenon? Mechanisms that keep neuronal firing within normal levels following alterations of visual experience, have been proposed to underlay the delayed increase of the open eye response (Mrsic-Flogel et al., 2007).

Two processes may compensate for the decrease of neural firing caused by eyelid suture: (i) a reduction in the threshold for synaptic potentiation, which increases the probability for strengthening of synaptic inputs (Abraham, 2008; Smith et al., 2009) and (ii) a scaling-up of synaptic efficacy that potentiates incoming drives (Turrigiano and Nelson, 2004). Work conducted in the laboratory of Mark Hubener and Tobias Bonhoeffer using two photon calcium imaging to quantify neuronal activity, has clearly demonstrated that homeostatic response regulation contributes to changes of eye-specific responsiveness after monocular deprivation in mouse visual cortex (Mrsic-Flogel et al., 2007).

Structural Changes Associated with Experience-Dependent Synaptic Plasticity

Recent *in vivo* imaging studies have demonstrated that experience-dependent modifications of visual cortex responsiveness are accompanied by an extensive structural remodeling of synaptic connectivity, as indicated by the growth and loss of dendritic spines (synaptic structures where neurons receive and integrate information). Dendritic spines in excitatory neurons, for instance, are particularly sensitive to experience as indicated by the observation that total lack of visual experience in early life (dark rearing) induces marked modifications in spine morphology and density, which are partially reversible by subsequent light exposure (Wallace and Bear, 2004). This notion is supported by the fact that monocular deprivation causes alterations in the motility, turnover, number and morphology of dendritic spines in the visual cortex (Majewska and Sur, 2003; Mataga et al., 2004; Hofer et al., 2009; Tropea et al., 2010).

To what extent does this form of structural plasticity contribute to experience-dependent changes of neural connectivity? Longitudinal two-photon imaging studies in the rodent visual system are beginning to address this question (Holmaat and Sbovoda, 2009). Recent *in vivo* studies on structural plasticity clearly indicate that dendritic spine dynamics is maximal during early stages of development but decreases thereafter, in parallel to the time-course of the critical period for visual cortex plasticity (Fu and Zuo, 2011). This is interesting, as it suggests that despite the absence of large-scale remodeling of dendrites, rearrangements of cortical connections in terms of growth and loss of spines may represent a structural substrate for experience-dependent plasticity. In line with this, it was recently observed that increasing the density and dynamics of dendritic spines by intracortical infusion of the bacterial toxin CNF1, is an effective strategy to restore plasticity in adult visual cortex (Cerri et al., 2011).

New synapse formation may increase memory storage capacity of the brain and new spines may even serve as structural traces for earlier memories, enabling the brain for faster adaptations to similar future experiences (Hofer et al., 2009; Xu et al., 2009). Experiments performed using the monocular deprivation paradigm seem to confirm this notion. Spine changes caused by a first experience of unilateral eye occlusion persist even after restoration of binocular vision and may therefore play a role in the enhancement of plasticity observed after a second episode of visual deprivation (Hofer et al., 2009).

The imaging studies mentioned above raise the question of whether structural modifications of dendritic spines represent functional changes of synaptic transmission. A recent electrophysiological study in hippocampal slice cultures points toward active functional properties of new spines, as indicated by the observation that a few hours after spine growth, AMPA- and NMDA-type glutamate receptor currents of new-born spines are similar to those of mature synaptic contacts (Zito et al., 2009).

Epigenetic Regulation of Adult Visual Cortical Plasticity

Neuronal plasticity is achieved by a complex interaction of molecular mechanisms, whereby intracellular signal pathways directly regulate gene transcription. In addition to the remarkable variety of transcription factors and their combinatorial interaction at specific gene promoters, the role of chromatin remodeling in regulating visual cortex plasticity has been increasingly appreciated.

Growing experimental evidence indicates that chromatin structure, including histone post-translational modifications, is highly dynamic within the nervous system and suggests the possibility that chromatin remodeling itself might be recruited as a target of plasticity-associated signal transduction mechanisms. Histone modifications, in fact, are thought to be actively involved in activity-dependent neural plasticity via regulation of gene expression.

Chromatin-Modifying Processes Restore Adult Visual Cortical Plasticity

Recent experimental evidence has shed light on molecular mechanisms that lie behind functional modifications in the visual system. Phenomena of plasticity during the critical period in the visual cortex of cats and rodents require the activation of intracellular protein kinases (e.g., PKA, ERK1/2,

CaMKII). The activation of intracellular second messenger pathways mediates alterations of chromatin structure, which in turn promote the expression of genes that allow plasticity to occur in different brain regions(Crosio et al., 2003; Levenson et al., 2004), this phenomenon being particularly prominent in early life. This is consistent with the recent observation that a developmental down-regulation of histone post-translational modifications control critical period plasticity in the mouse visual system (Putignano et al., 2007). In line with this notion, a pharmacologically-induced modification of chromatin structure restores plasticity in the adult visual cortex, as indicated by the fact that directly increasing acetylation of histones by means of the histone deacetylase inhibitor: Trichostatin-A effectively reactivates susceptibility to monocular deprivation in rodents (Putignano et al., 2007; Maya-Vetencourt et al., 2011).

Can experience trigger epigenetic mechanisms of plasticity in the adult? Insights intothis initially came from the observation that ascending projection systems, the levels of which vary in response to experience, play a key role in mediating plasticity in the kitten visual cortex (Kasamatsu and Pettigrew, 1976; Bear and Singer, 1986; Gu and Singer, 1995; Wang et al., 1997). This notion has been recently confirmed by two recent observations. Firstly, a gradual increase in the expression of the Linx1 protein prevents plasticity in the developing visual cortex and the genetic removal of this protein restores plasticity in adulthood by enhancing nicotinic acetylcholine receptor signaling (Morishita et al., 2010). Secondly, direct infusion of serotonin into the adult visual cortex not only reinstates juvenile-like plasticity in the rodent visual system,but promote modifications of chromatin structure associated withgene transcription activation (Maya-Vetencourt et al., 2011). Moreover, an increased serotonergic signaling is a common finding in environmental enrichment and chronic fluoxetine administration, two different experimental strategies which reinstate a degree of plasticity in the visual system that is similar to that observed during the critical period and promote the recovery of visual functions after long-term sensory deprivation (Sale et al., 2007; Baroncelli et al., 2010; Maya-Vetencourt et al., 2008; 2011). Importantly, these two experimental approaches consistently alter the inhibitory/excitatory balance in the visual cortex while promoting BDNF expression and LTP occurrence(Maya-Vetencourt et al., 2008; Baroncelli et al., 2010).

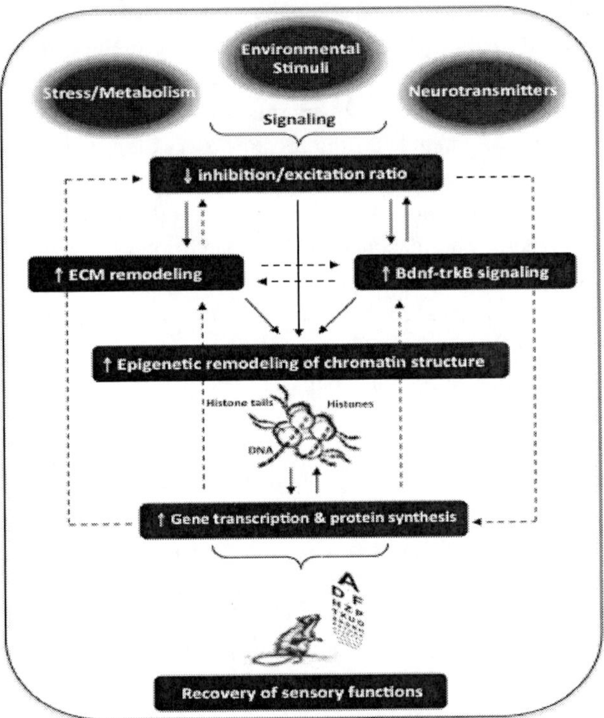

Figure 2. The process of plasticity reactivation is associated withsignal transduction pathways that involve the activation of ascending projection systems. Serotonergic and cholinergic transmission set in motion physiological processes that underlie plasticity in adult life by reducing the inhibitory/excitatory ratio (Morishita et al., 2010; Baroncelli et al., 2010; Maya-Vetencourt et al., 2008; 2011). A reduction of the inhibitory/excitatory balance may directly activate intracellular mechanisms that eventually promote epigenetic modifications of chromatin structure (e.g., post-translational modifications of histones), which in turn allow for the expression of genes that act as down-stream effectors of plastic phenomena in adult life. A pharmacological shift of the inhibitory/excitatory ratio does, in fact, promote the activity-dependent *BDNF* expression and degradation of extracellular matrix components that are inhibitory for plasticity (Harauzov et al., 2010). BDNF-trkB signaling might up-regulate additional gene expression patterns associated with functional modifications in the visual cortex. This could also alter the balance of intracortical inhibition and excitation (unpublished data). Degradation of extracellular matrix components may modify the inhibition/excitation ratio in the visual system. The interaction between BDNF-trkB signaling and extracellular matrix reorganization has yet to be explored. Continuous arrows represent established interactions between the molecular and cellular processes mentioned (boxes). Dashed lines represent interactions that remain to be ascertained.

Among the 14 known 5-HT receptors, the 5-HT_{1A} subtype seem to mediate these plastic phenomena as indicated by the fact that intracortical infusion of 5-HT_{1A} antagonists in fluoxetine-treated animals effectively prevent the process of plasticity reactivation. The 5-HT_{1A}-dependent enhancement of plasticity is likely to be mediated by a shift of the intracortical inhibitory/excitatory balance, as it has been reported that 5-HT_{1A} receptors are expressed in diverse types of GABAergic interneurons (Aznar et al., 2003) and down-regulate GABAergic transmission in different regions of the brain (Schmitz et al., 1995; Koyama et al., 1999; Xiang and Prince, 2003; Bramley et al., 2005).

Moreover, 5-HT_{1A} receptors are expressed in parvalbumin-positive GABAergic interneurons in the adult visual cortex and cortical infusion of agonists of this receptor subtype restores susceptibility to monocular deprivation, an event that is paralleled by a reduction of extracellular levels of GABA (Maya-Vetencourt et al., unpublished data).This may decrease the threshold for visual cortical neurons to be driven by electrical activity, thus increasing the probability for visual inputs to drive activity-dependent modifications of synaptic transmission. These findings suggest a scenario in whichan enhanced serotonergic and/or cholinergic signaling triggers an initial cascade of molecular and cellular events that eventually converge to promote changes of chromatin structure that underlie gene expression profiles associated with functional modifications in the adult visual system (see Figure 2). The reinstatement of plasticity caused by fluoxetine in the rodent visual systemis, in fact, accompanied by a transitory increment of *BDNF* expression that is paralleled by an enhanced histone acetylation status at the activity-dependently regulated *BDNF* promoter regions of exons I and IV and by a decreased expression of histone deacetylase (HDAC) enzymes (Maya-Vetencourt et al., 2011). In keeping with this, long-term treatment with HDACs inhibitors (e.g., valproic acid and sodium butyrate) promotes a complete recovery of visual functions in adult amblyopic animals (Silingardi et al., 2010).

Conclusion

Although the ability of the nervous system to change in response to experience is particularly high during early life,years of selective pressure through which the brain has evolved have ensured modifications of brain functions to changing environmental conditions during the entire lifespan. In fact, the reorganization of cortical circuitries in adulthood persists, at least, under some circumstances. These findings highlight the fact that achieving a fundamental understanding of physiological processes that control neuronal plasticity may be of clinical relevance in a wide range of neurological disorders in which a rearrangement of neural circuitries in adult life may be beneficial.

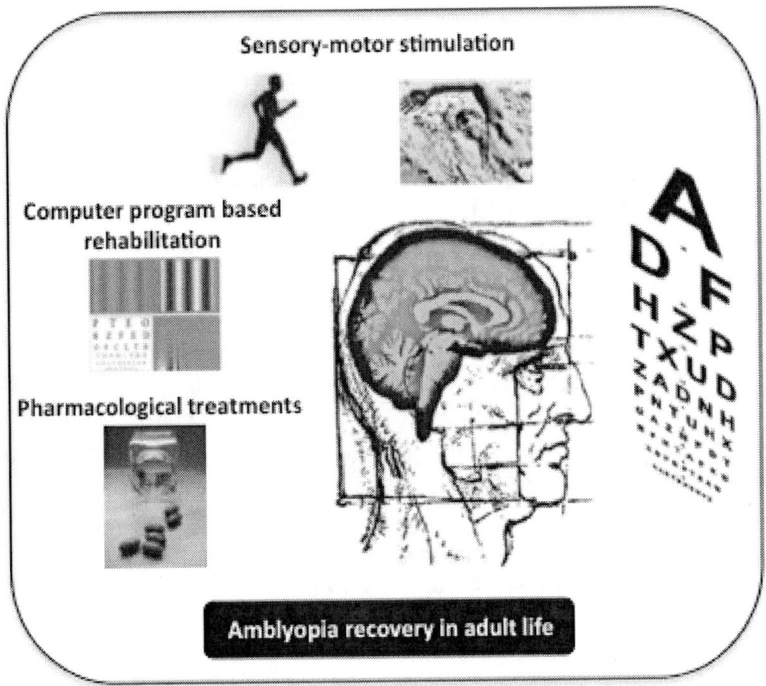

Figure 3. Potential strategies for the treatment of human amblyopia. The recent findings that environmental enrichment (Sale et al., 2007; Baroncelli et al., 2010), chronic administration of fluoxetine (Maya-Vetencourt et al., 2008; 2011), dark exposure (He et al., 2006) and food restriction (Spolidoro et al., 2011) all promote full recovery of visual acuity and binocularity in adult amblyopic animals, both at electrophysiological and behavioral level, emphasize the clinical potential of pharmacological strategies that modulate serotonergic, cholinergic and/or GABAergic transmission, together with an enhanced sensory-motor activity in human amblyopia. In this context, an enhanced physical activity (e.g., running, swimming, other sensory modalities such as tactile stimulation -body massage-), and pharmacological strategiesmay be complementary to perceptual learning in the treatment of amblyopia after 9 years of age, after which time this pathology is normally refractory to treatment.

The process of plasticity reactivation in the visual system seems to be a multi-factorial mechanism that includes at least four important events: (i) an increased serotonergic and/or cholinergic signaling, (ii) a shift of the intracortical inhibitory/excitatory balance, (iii) an enhancedneurotrophin expression and signaling, as well as (iv) modifications at the level of chromatin structure that allow for the expression of genes that mediate structural and functional changes in the nervous system. The potential clinical application of strategies that promote adult plasticity in the visual cortexhas long been explored with the prospect of treating human amblyopia, a pathological condition that arises from an abnormal visual experience during development and is refractory treatment in the adult (Holmes and Clark, 2006).

The data on animal models previously reported suggests that an enhanced sensory-motor activity (Sale et al., 2007; Baroncelli et al., 2010), food restriction (Spolidoro et al., 2011) and/or chronic fluoxetine administration (Maya-Vetencourt et al., 2008; 2011), are non-invasive experimental strategies that may be used as complementary treatments to current therapies for human amblyopia (see Figure 3).

In this context, perceptual learning has long been used to improve spatial acuity in adult amblyopic patients (Levi and Li, 2009). Systematic training of patients with unilateral amblyopia (secondary to strabismus and anisometropia) in simple visual tasks revealed a 2-fold increase of contrast sensitivity and improved performance in letter-recognition tests (Polat et al., 2004). Likewise, Snellen acuities in anisometric amblyopes improved after intensive training in a Vernier acuity task. The improvement of performance seen in perceptual learning is proportional to the number of trials taken, although performance eventually reaches an asymptote of no further progress (Huang et al., 2008). Unfortunately, the extent to which acuity improvements occur is limited by the task specificity of perceptual learning (Crist et al., 1997).

In most instances of perceptual learning, attention to the trained stimulus is necessary for improvements of vision to occur (Ahissar and Hochstein, 1993). This observation is particularly important as it epitomizes the role of ascending projection systems in regulating mechanisms of attention and information processing that may contribute to functional modifications in the adult brain. These findings portray the enhancement of plasticity as a strategy for brain repair in adult life; particularly, in a variety of pathological states where reorganization of neuronal networks would be beneficial in adulthood. This could facilitate the restructuring of mature circuitries impaired by damage or disease. Indeed, long-term FLX administration, which increases serotonergic transmission, enhances the effects of rehabilitation in the recovery from motor deficits after ischaemic stroke in humans (Chollet et al., 2011). These observations emphasize the fact that neuronal representations in adulthood are dynamic and can be continuously modified by sensory experience.

REFERENCES

Abraham, W.C. (2008). Metaplasticity: tuning synapses and networks for plasticity. *Nature reviews Neuroscience* 9:387-399.

Ahissar, M. and Hochstein, S. (1993). Attentional control of early perceptual learning. Proceedings of the National Academy of Sciences of the United States of America 90:5718-5722.

Atwal, J.K., Pinkston-Gosse, J., Syken, J., Stawicki, S., Wu, Y., Shatz, C., Tessier-Lavigne, M. (2008). PirB is a functional receptor for myelin inhibitors of axonal regeneration. *Science* 322:967-970.

Aznar, S., Qian, Z., Shah, R., Rahbek, B., Knudsen, G.M. (2003).The 5-HT1A serotonin receptor is located on calbindin- and parvalbumin-containing neurons in the rat brain. *Brain research* 959:58-67.

Bansal, A., Singer, J.H., Hwang, B.J., Xu, W., Beaudet, A., Feller, M.B. (2000). Mice lacking specific nicotinic acetylcholine receptor subunits exhibit dramatically altered spontaneous activity patterns and reveal a limited role for retinal waves in forming ON and OFF circuits in the inner retina. *The Journal of neuroscience* 20:7672-7681.

Baroncelli, L., Sale, A., Viegi, A., Maya-Vetencourt, J.F., De Pasquale, R., Baldini, S., Maffei, L. (2010).Experience-dependent reactivation of ocular dominance plasticity in the adult visual cortex.*Experimental Neurology*226:100-109.

Bear, M.F., Kleinschmidt, A., Gu, Q.A., Singer, W. (1990). Disruption of experience-dependent synaptic modifications in striate cortex by infusion of an NMDA receptor antagonist.*The Journal of neuroscience* 10:909-925.

Bear, M.F. and Singer, W. (1986) Modulation of visual cortical plasticity by acetylcholine and noradrenaline. *Nature* 320:172-176.

Beaver, C.J., Ji, Q., Fischer, Q.S., Daw, N.W. (2001). Cyclic AMP-dependent protein kinase mediates ocular dominance shifts in cat visual cortex. *Nature neuroscience* 4:159-163.

Berardi, N., Cellerino, A., Domenici, L., Fagiolini, M., Pizzorusso, T., Cattaneo, A., Maffei, L. (1994). Monoclonal antibodies to nerve growth factor affect the postnatal development of the visual system. Proceedings of the National Academy of Sciences of the United States of America 91:684-688.

Berardi, N., Pizzorusso, T., Maffei, L. (2000).Critical periods during sensory development.*Current opinion in neurobiology* 10:138-145.

Betts, L.R., Taylor, C.P., Sekuler, A.B., Bennett, P.J. (2005). Aging reduces center-surround antagonism in visual motion processing. *Neuron* 45:361-366.

Bienenstock, E.L., Cooper, L.N., Munro, P.W. (1982). Theory for the development of neuron selectivity: orientation specificity and binocular interaction in visual cortex. *The Journal of neuroscience* 2:32-48.

Bramley, J.R., Sollars, P.J., Pickard, G.E., Dudek, F.E. (2005). 5-HT1B receptor-mediated presynaptic inhibition of GABA release in the suprachiasmatic nucleus.*Journal of neurophysiology* 93:3157-3164.

Cabelli, R.J., Hohn, A., Shatz, C.J. (1995). Inhibition of ocular dominance column formation by infusion of NT-4/5 or BDNF.*Science* 267:1662-1666.

Cabelli, R.J., Shelton, D.L., Segal, R.A., Shatz, C.J. (1997). Blockade of endogenous ligands of trkB inhibits formation of ocular dominance columns. *Neuron* 19:63-76.

Caleo, M., Lodovichi, C., Maffei, L. (1999). Effects of nerve growth factor on visual cortical plasticity require afferent electrical activity. *The European journal of neuroscience* 11:2979-2984.

Carmignoto, G., Canella, R., Candeo, P., Comelli, M.C., Maffei, L. (1993). Effects of nerve growth factor on neuronal plasticity of the kitten visual cortex. *The Journal of physiology* 464:343-360.

Carmignoto, G., Vicini, S. (1992). Activity-dependent decrease in NMDA receptor responses during development of the visual cortex.*Science* 258:1007-1011.

Cerri, C., Fabbri, A., Vannini, A., Spolidoro, M., Costa, M., Maffei, L., Fiorentini, C., Caleo, M. (2011).Activation of Rho GTPases triggers structural remodeling and functional plasticity in the adult rat visual cortex. *Journal of Neuroscience*31:15163-15172.

Chen, J.L., Lin, W.C., Cha, J.W., So, P.T., Kubota, Y., Nedivi, E. (2011). Structural basis for the role of inhibition in facilitating adult brain plasticity.*Nature Neuroscience*14:587-594.

Chen, L., Cooper, N.G., Mower, G.D. (2000). Developmental changes in the expression of NMDA receptor subunits (NR1, NR2A, NR2B) in the cat visual cortex and the effects of dark rearing.*Brain research Molecular brain research* 78:196-200.

Cho, K.K., Khibnik, L., Philpot, B.D., Bear, M.F. (2009). The ratio of NR2A/B NMDA receptor subunits determines the qualities of ocular dominance plasticity in visual cortex. Proceedings of the National Academy of Sciences of the United States of America 106:5377-5382.

Chollet, F., Tardy, J., Albucher, J.F., Thalamas, C., Berard, E., Lamy, C., Bejot, Y., Deltour, S., Jaillard, A., Niclot, P., Guillon, B., Moulin, T., Marque, P., Pariente, J., Arnaud, C., Loubinoux, I. (2011). Fluoxetine for motor recovery after acute ischaemic stroke (FLAME): a randomised placebo-controlled trial. *Lancet neurology* 10:123-130.

Crist, R.E., Kapadia, M.K., Westheimer, G., Gilbert, C.D. (1997). Perceptual learning of spatial localization: specificity for orientation, position, and context. *Journal of neurophysiology* 78:2889-2894.

Crosio, C., Heitz, E., Allis, C.D., Borrelli, E., Sassone-Corsi, P. (2003). Chromatin remodeling and neuronal response: multiple signaling pathways induce specific histone H3 modifications and early gene expression in hippocampal neurons. *Journal of cell science* 116:4905-4914.

Di Cristo, G., Berardi, N., Cancedda, L., Pizzorusso, T., Putignano, E., Ratto, G.M., Maffei, L. (2001).Requirement of ERK activation for visual cortical plasticity.*Science* 292:2337-2340.

Domenici, L., Berardi, N., Carmignoto, G., Vantini, G., Maffei, L. (1991). Nerve growth factor prevents the amblyopic effects of monocular deprivation. Proceedings of the National Academy of Sciences of the United States of America 88:8811-8815.

Domenici, L., Cellerino, A., Berardi, N., Cattaneo, A., Maffei, L. (1994). Antibodies to nerve growth factor (NGF) prolong the sensitive period for monocular deprivation in the rat. *Neuroreport*5:2041-2044.

Fagiolini, M., Fritschy, J.M., Low, K., Mohler, H., Rudolph, U., Hensch, T.K. (2004). Specific GABAA circuits for visual cortical plasticity.*Science* 303:1681-1683.

Fagiolini, M. and Hensch, T.K. (2000). Inhibitory threshold for critical-period activation in primary visual cortex.*Nature* 404:183-186.

Fagiolini, M., Katagiri, H., Miyamoto, H., Mori, H., Grant, S.G., Mishina, M., Hensch, T.K. (2003). Separable features of visual cortical plasticity revealed by N-methyl-D-aspartate receptor 2A signaling. Proceedings of the National Academy of Sciences of the United States of America 100:2854-2859.

Fagiolini, M., Pizzorusso, T., Berardi, N., Domenici, L., Maffei, L. (1994). Functional postnatal development of the rat primary visual cortex and the role of visual experience: dark rearing and monocular deprivation. *Vision research* 34:709-720.

Flint, A.C., Maisch, U.S., Weishaupt, J.H., Kriegstein, A.R., Monyer, H. (1997). NR2A subunit expression shortens NMDA receptor synaptic currents in developing neocortex. *The Journal of neuroscience* 17:2469-2476.

Frenkel, M.Y. and Bear, M.F. (2004). How monocular deprivation shifts ocular dominance in visual cortex of young mice.*Neuron* 44:917-923.

Fu, M. and Zuo, Y. (2011). Experience-dependent structural plasticity in the cortex.*Trends in neurosciences* 34:177-187.

Galli, L. and Maffei, L. (1988). Spontaneous impulse activity of rat retinal ganglion cells in prenatal life. *Science* 242:90-91.

Galuske, R.A., Kim, D.S., Castren, E., Thoenen, H., Singer, W. (1996). Brain-derived neurotrophic factor reversed experience-dependent synaptic modifications in kitten visual cortex. *The European journal of neuroscience* 8:1554-1559.

Gordon, J.A., Cioffi, D., Silva, A.J., Stryker, M.P. (1996). Deficient plasticity in the primary visual cortex of alpha-calcium/calmodulin-dependent protein kinase II mutant mice.*Neuron* 17:491-499.

Gu, Q. and Singer, W. (1995). Involvement of serotonin in developmental plasticity of kitten visual cortex.*The European journal of neuroscience* 7:1146-1153.

Harauzov, A., Spolidoro, M., Di Cristo, G., De Pasquale, R., Cancedda, L., Pizzorusso, T., Viegi, A., Berardi, N., Maffei, L. (2010). Reducing intracortical inhibition in the adult visual cortex promotes ocular dominance plasticity. *The Journal of neuroscience* 30:361-371.

He, H.Y., Hodos, W., Quinlan, E.M. (2006). Visual deprivation reactivates rapid ocular dominance plasticity in adult visual cortex. *The Journal of neuroscience* 26:2951-2955.

Hensch, T.K. (2005). Critical period plasticity in local cortical circuits.*Nature reviews Neuroscience* 6:877-888.

Hensch, T.K., Fagiolini, M., Mataga, N., Stryker, M.P., Baekkeskov, S., Kash, S.F. (1998). Local GABA circuit control of experience-dependent plasticity in developing visual cortex.*Science* 282:1504-1508.

Heynen, A.J., Yoon, B.J., Liu, C.H., Chung, H.J., Huganir, R.L., Bear, M.F. (2003). Molecular mechanism for loss of visual cortical responsiveness following brief monocular deprivation.*Nature neuroscience* 6:854-862.

Hofer, S.B., Mrsic-Flogel, T.D., Bonhoeffer, T., Hubener, M. (2009). Experience leaves a lasting structural trace in cortical circuits. *Nature* 457:313-317.

Holmes, J.M. and Clarke, M.P. (2006). Amblyopia. *Lancet* 367:1343-1351.

Holtmaat, A. and Svoboda, K. (2009). Experience-dependent structural synaptic plasticity in the mammalian brain.*Nature reviews Neuroscience* 10:647-658.

Huang, C.B., Zhou, Y., Lu, Z.L. (2008). Broad bandwidth of perceptual learning in the visual system of adults with anisometropic amblyopia. Proceedings of the National Academy of Sciences of the United States of America 105:4068-4073.

Huang, Z.J., Kirkwood, A., Pizzorusso, T., Porciatti, V., Morales, B., Bear, M.F., Maffei, L., Tonegawa, S. (1999). BDNF regulates the maturation of inhibition and the critical period of plasticity in mouse visual cortex. *Cell* 98:739-755.,

Kasamatsu, T. and Pettigrew, J.D. (1976). Depletion of brain catecholamines: failure of ocular dominance shift after monocular occlusion in kittens. *Science* 194:206-209.

Katz, L.C. and Shatz, C.J. (1996). Synaptic activity and the construction of cortical circuits.*Science* 274:1133-1138.

Kirkwood, A., Dudek, S.M., Gold, J.T., Aizenman, C.D., Bear, M.F. (1993). Common forms of synaptic plasticity in the hippocampus and neocortex in vitro.*Science* 260:1518-1521.

Koyama, S., Kubo, C., Rhee, J.S., Akaike, N. (1999). Presynaptic serotonergic inhibition of GABAergic synaptic transmission in mechanically dissociated rat basolateral amygdala neurons. *The Journal of physiology* 518 (2):525-538.

LeVay, S., Wiesel, T.N., Hubel, D.H. (1980).The development of ocular dominance columns in normal and visually deprived monkeys.*The Journal of comparative neurology* 191:1-51.

Levenson, J.M., O'Riordan, K.J., Brown, K.D., Trinh, M.A., Molfese, D.L., Sweatt, J.D. (2004). Regulation of histone acetylation during memory formation in the hippocampus.*The Journal of biological chemistry* 279:40545-40559.

Leventhal, A.G., Wang, Y., Pu, M., Zhou, Y., Ma, Y. (2003). GABA and its agonists improved visual cortical function in senescent monkeys. *Science* 300:812-815.

Levi, D.M., Li, R.W. (2009). Perceptual learning as a potential treatment for amblyopia: a mini-review. *Vision research* 49:2535-2549.

Lodovichi, C., Berardi, N., Pizzorusso, T., Maffei, L. (2000). Effects of neurotrophins on cortical plasticity: same or different? *The Journal of neuroscience* 20:2155-2165.

Lu, H., Cheng, P.L., Lim, B.K., Khoshnevisrad, N., Poo, M.M. (2010). Elevated BDNF after cocaine withdrawal facilitates LTP in medial prefrontal cortex by suppressing GABA inhibition. *Neuron* 67:821-833.

Maffei, L., Berardi, N., Domenici, L., Parisi, V., Pizzorusso, T. (1992). Nerve growth factor (NGF) prevents the shift in ocular dominance distribution of visual cortical neurons in monocularly deprived rats. *The Journal of neuroscience* 12:4651-4662.

Majewska, A., Sur, M. (2003). Motility of dendritic spines in visual cortex in vivo: changes during the critical period and effects of visual deprivation. Proceedings of the National Academy of Sciences of the United States of America 100:16024-16029.

Mataga, N., Imamura, K., Shiomitsu, T., Yoshimura, Y., Fukamauchi, K., Watanabe, Y. (1996). Enhancement of mRNA expression of tissue-type plasminogen activator by L-threo-3,4-dihydroxyphenylserine in association with ocular dominance plasticity. *Neuroscience letters* 218:149-152.

Mataga, N., Mizuguchi, Y., Hensch, T.K. (2004). Experience-dependent pruning of dendritic spines in visual cortex by tissue plasminogen activator.*Neuron* 44:1031-1041.

Mataga, N., Nagai, N., Hensch, T.K. (2002). Permissive proteolytic activity for visual cortical plasticity. Proceedings of the National Academy of Sciences of the United States of America 99:7717-7721.

Maya-Vetencourt, J.F., Sale, A., Viegi, A., Baroncelli, L., De Pasquale, R., O'Leary, O.F., Castren, E., Maffei, L. (2008). The antidepressant fluoxetine restores plasticity in the adult visual cortex. *Science* 320:385-388.

Maya-Vetencourt, J.F., Tiraboschi, E., Spolidoro, M., Castren, E., Maffei, L. (2011). Serotonin triggers a transient epigenetic mechanism that reinstates adult visual cortex plasticity in rats. *The European journal of neuroscience* 33:49-57.

McGee, A.W., Yang, Y., Fischer, Q.S., Daw, N.W., Strittmatter, S.M. (2005). Experience-driven plasticity of visual cortex limited by myelin and Nogo receptor. *Science* 309:2222-2226.

Meister, M., Wong, R.O., Baylor, D.A., Shatz, C.J. (1991). Synchronous bursts of action potentials in ganglion cells of the developing mammalian retina. *Science* 252:939-943.

Morishita, H., Miwa, J.M., Heintz, N., Hensch, T.K. (2010). Lynx1, a cholinergic brake, limits plasticity in adult visual cortex. *Science* 330:1238-1240.

Mower, A.F., Liao, D.S., Nestler, E.J., Neve, R.L., Ramoa, A.S. (2002).cAMP/Ca2+ response element-binding protein function is essential for ocular dominance plasticity. *The Journal of neuroscience* 22:2237-2245.

Mrsic-Flogel, T.D., Hofer, S.B., Ohki, K., Reid, R.C., Bonhoeffer, T., Hubener, M. (2007).Homeostatic regulation of eye-specific responses in visual cortex during ocular dominance plasticity.*Neuron* 54:961-972.

Muller, C.M. and Griesinger, C.B. (1998). Tissue plasminogen activator mediates reverse occlusion plasticity in visual cortex. *Nature neuroscience* 1:47-53.

Oray, S., Majewska, A., Sur, M. (2004). Dendritic spine dynamics are regulated by monocular deprivation and extracellular matrix degradation. *Neuron* 44:1021-1030.

Pham, T.A., Impey, S., Storm, D.R., Stryker, M.P. (1999). CRE-mediated gene transcription in neocortical neuronal plasticity during the developmental critical period. *Neuron* 22:63-72.

Pizzorusso, T., Berardi, N., Rossi, F.M., Viegi, A., Venstrom, K., Reichardt, L.F., Maffei, L. (1999).TrkA activation in the rat visual cortex by antirattrkAIgG prevents the effect of monocular deprivation. *The European journal of neuroscience* 11:204-212.

Pizzorusso, T., Medini, P., Berardi, N., Chierzi, S., Fawcett, J.W., Maffei, L. (2002).Reactivation of ocular dominance plasticity in the adult visual cortex.*Science* 298:1248-1251.

Pizzorusso, T., Medini, P., Landi, S., Baldini, S., Berardi, N., Maffei, L. (2006).Structural and functional recovery from early monocular deprivation in adult rats. Proceedings of the National Academy of Sciences of the United States of America 103:8517-8522.

Polat, U., Ma-Naim, T., Belkin, M., Sagi, D. (2004). Improving vision in adult amblyopia by perceptual learning. Proceedings of the National Academy of Sciences of the United States of America 101:6692-6697.

Putignano, E., Lonetti, G., Cancedda, L., Ratto, G., Costa, M., Maffei, L., Pizzorusso, T. (2007). Developmental downregulation of histone posttranslational modifications regulates visual cortical plasticity. *Neuron* 53:747-759.

Quinlan, E.M., Philpot, B.D., Huganir, R.L., Bear, M.F. (1999). Rapid, experience-dependent expression of synaptic NMDA receptors in visual cortex in vivo.*Nature neuroscience* 2:352-357.

Roberts, E.B., Meredith, M.A., Ramoa, A.S. (1998). Suppression of NMDA receptor function using antisense DNA block ocular dominance plasticity while preserving visual responses.*Journal of neurophysiology* 80:1021-1032.

Sale, A., Maya-Vetencourt, J.F., Medini, P., Cenni, M.C., Baroncelli, L., De Pasquale, R., Maffei, L. (2007). Environmental enrichment in adulthood promotes amblyopia recovery through a reduction of intracortical inhibition. *Nature neuroscience* 10:679-681.

Schmitz, D., Empson, R.M., Heinemann, U. (1995). Serotonin reduces inhibition via 5-HT1A receptors in area CA1 of rat hippocampal slices in vitro. *The Journal of neuroscience* 15:7217-7225.

Shatz, C.J. and Stryker, M.P. (1988). Prenatal tetrodotoxin infusion blocks segregation of retinogeniculate afferents. *Science* 242:87-89.

Silingardi, D., Scali, M., Belluomini, G., Pizzorusso, T. (2010). Epigenetic treatments of adult rats promote recovery from visual acuity deficits induced by long-term monocular deprivation. *The European journal of neuroscience* 31:2185-2192.

Smith, G.B., Heynen, A.J., Bear, M.F. (2009). Bidirectional synaptic mechanisms of ocular dominance plasticity in visual cortex.Philosophical transactions of the Royal Society of London Series B, *Biological sciences* 364:357-367.

Snider, C.J., Dehay, C., Berland, M., Kennedy, H., Chalupa, L.M. (1999).Prenatal development of retinogeniculate axons in the macaque monkey during segregation of binocular inputs.*The Journal of neuroscience* 19:220-228.

Spolidoro, M., Baroncelli, L., Putignano, E., Maya-Vetencourt, J.F., Viegi, A., Maffei, L. (2011). Food restriction enhances visual cortex plasticity in adulthood. *Nature communications* 2:320.

Sugiyama, S., Di Nardo, A.A., Aizawa, S., Matsuo, I., Volovitch, M., Prochiantz, A., Hensch, T.K. (2008). Experience-dependent transfer of Otx2 homeoprotein into the visual cortex activates postnatal plasticity. *Cell* 134:508-520.

Syken, J., Grandpre, T., Kanold, P.O., Shatz, C.J. (2006).PirB restricts ocular-dominance plasticity in visual cortex. *Science* 313:1795-1800.

Taha, S., Hanover, J.L., Silva, A.J., Stryker, M.P. (2002). Autophosphorylation of alphaCaMKII is required for ocular dominance plasticity. *Neuron* 36:483-491.

Tropea, D., Majewska, A.K., Garcia, R., Sur, M. (2010). Structural dynamics of synapses in vivo correlate with functional changes during experience-dependent plasticity in visual cortex. *The Journal of neuroscience* 30:11086-11095.

Turrigiano, G.G. and Nelson, S.B. (2004). Homeostatic plasticity in the developing nervous system.*Nature reviews Neuroscience* 5:97-107.

Wallace, W. and Bear, M.F. (2004).A morphological correlate of synaptic scaling in visual cortex.*The Journal of neuroscience* 24:6928-6938.

Wang, Y., Gu, Q., Cynader, M.S. (1997). Blockade of serotonin-2C receptors by mesulergine reduces ocular dominance plasticity in kitten visual cortex. *Experimental brain research*114:321-328.

Wiesel, T.N. and Hubel, D.H. (1963). Effects of Visual Deprivation on Morphology and Physiology of Cells in the Cats Lateral Geniculate Body.*Journal of neurophysiology* 26:978-993.

Wiesel, T.N. and Hubel, D.H. (1963). Single-Cell Responses in Striate Cortex of Kittens Deprived of Vision in One Eye.*Journal of neurophysiology* 26:1003-1017.

Wiesel, T.N. and Hubel, D.H. (1974). Ordered arrangement of orientation columns in monkeys lacking visual experience. *The Journal of comparative neurology* 158:307-318.

Xiang, Z. and Prince, D.A. (2003). Heterogeneous actions of serotonin on interneurons in rat visual cortex.*Journal of neurophysiology* 89:1278-1287.

Xu, T., Yu, X., Perlik, A.J., Tobin, W.F., Zweig, J.A., Tennant, K., Jones, T., Zuo, Y. (2009).Rapid formation and selective stabilization of synapses for enduring motor memories.*Nature* 462:915-919.

Zito, K., Scheuss, V., Knott, G., Hill, T., Svoboda, K. (2009). Rapid functional maturation of nascent dendritic spines.*Neuron* 61:247-258.

In: Visual Cortex: Anatomy, Functions ... ISBN: 978-1-62100-948-1
Editors: J.M. Harris et al. pp. 99-128 © 2012 Nova Science Publishers, Inc.

Chapter 4

THE VISUAL CORTEX IN ALZHEIMER'S DISEASE: LAMINAR DISTRIBUTION OF THE PATHOLOGICAL CHANGES IN VISUAL AREAS V1 AND V2

Richard A. Armstrong*
Vision Sciences, Aston University,
Birmingham B4 7ET, UK.

ABSTRACT

Alzheimer's disease (AD) is an important neurodegenerative disorder causing visual problems in the elderly population. The pathology of AD includes the deposition in the brain of abnormal aggregates of β-amyloid (Aβ) in the form of senile plaques (SP) and abnormally phosphorylated tau in the form of neurofibrillary tangles (NFT). A variety of visual problems have been reported in patients with AD including loss of visual acuity (VA), colour vision and visual fields; changes in pupillary responses to mydriatics, defects in fixation and in smooth and saccadic eye movements; changes in contrast sensitivity and in visual evoked potentials (VEP); and disturbances in complex visual tasks such as reading, visuospatial function, and in the naming and identification of objects. In addition, pathological changes have been observed to affect

---

* Corresponding author: Dr. RA Armstrong, Vision Sciences, Aston University, Birmingham B4 7ET, UK. Tel: 0121-204-4102; Fax: 0121-204-3892; Email: R.A.Armstrong@aston.ac.uk

the eye, visual pathway, and visual cortex in AD. To better understand degeneration of the visual cortex in AD, the laminar distribution of the SP and NFT was studied in visual areas V1 and V2 in 18 cases of AD which varied in disease onset and duration. In area V1, the mean density of SP and NFT reached a maximum in lamina III and in laminae II and III respectively. In V2, mean SP density was maximal in laminae III and IV and NFT density in laminae II and III. The densities of SP in laminae I of V1 and NFT in lamina IV of V2 were negatively correlated with patient age. No significant correlations were observed in any cortical lamina between the density of NFT and disease onset or duration. However, in area V2, the densities of SP in lamina II and lamina V were negatively correlated with disease duration and disease onset respectively. In addition, there were several positive correlations between the densities of SP and NFT in V1 with those in area V2. The data suggest: (1) NFT pathology is greater in area V2 than V1, (2) laminae II/III of V1 and V2 are most affected by the pathology, (3) the formation of SP and NFT in V1 and V2 are interconnected, and (4) the pathology may spread between visual areas via the feed-forward short cortico-cortical connections.

Keywords: Alzheimer's disease (AD), visual cortex, striate cortex (B17), extrastriate cortex (B18), senile plaques, neurofibrillary tangles (NFT).

INTRODUCTION

Alzheimer's disease (AD) is a degenerative disorder of the nervous system affecting approximately 10% of individuals aged 65 or over (Knopman, 2001). It is estimated that 24.3 million individuals worldwide may have dementia and 4.6 million new cases are recorded every year (Ferri et al., 2005). There have been various estimates of the prevalence of AD in the elderly population (Evans et al., 1989; Ferri et al., 2005). With advancing age, the prevalence of the disease increases to an estimated 19% in individuals 75-84 years (Knopman, 2001), and is 30-35% for those older than 85 years (Ferri et al., 2005). The development of dementia involves a decline in short-term memory, impairment of judgment, and a loss of emotional control. These symptoms develop slowly over a period of years resulting ultimately in a complete breakdown of mental function. A small proportion of cases of AD (<5%) have a genetic origin but most cases occur sporadically within the population (Hoenicka, 2006).

The clinical diagnosis of AD is based on criteria developed originally by the 'National Institute of Neurological and Communicative Disorders and

Stroke and the Alzheimer's Disease and Related Disorders Association' (NINCDS-ADRDA) work group (Tierney et al., 1988) and modified by the National Institute on Aging (NIA) - Reagan Institute (Jellinger and Bancher, 1998). The definitive diagnosis of AD, however, requires examination of brain tissue either at biopsy or post-mortem. The pathology of AD includes the deposition in the brain of abnormal aggregates of β-amyloid (Aβ) in the form of senile plaques (SP) and abnormally phosphorylated tau in the form of neurofibrillary tangles (NFT). Hence, significant densities of SP and NFT in the cerebral cortex of the brain are required for a pathological diagnosis of AD. There are no treatments that can arrest the progression of the disease but some drugs can successfully treat the symptoms for a period of time.

PATHOLOGY OF ALZHEIMER'S DISEASE (AD)

General Features

AD is characterized by a large number of pathological changes affecting the brain, many of which appear to be exacerbations of normal aging (Armstrong, 2008). There is atrophy of the cerebral hemispheres with narrowed gyri and widened sulci and widening of the subarachnoid space (Imhof et al., 2007). These gross changes mainly affect the frontal, temporal, and parietal lobes and often spare the occipital lobe. The meninges are thickened by fibrosis and the ventricles are often dilated (Brun, 1983). The white matter appears yellow and 'rubbery' and there is 'spongiosis' and 'gliosis' affecting the grey matter (Brun, 1983). These changes may be more apparent in cases of early-onset (<65 years) while some late-onset cases (>65 years) may exhibit little atrophy (Mortimer, 1983).

Histological Features

Ever since the first descriptions of pre-senile dementia by Alois Alzheimer in 1907 (Alzheimer, 1907), the formation of SP and NFT (Figure 1) have been regarded as the defining histological features of Alzheimer's disease (AD) (Khatchaturian, 1985; Mirra et al., 1991; Jellinger and Bancher, 1998). The most important molecular constituent of the SP is β-amyloid (Aβ) (Glenner and Wong, 1984) and consequently SP are also referred to as Aβ deposits.

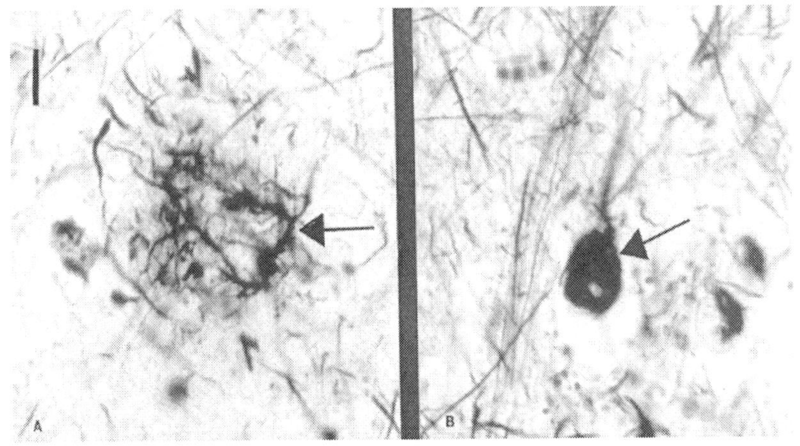

Figure 1. Hallmark lesions of Alzheimer's disease: (A) senile plaque (SP) and (B) neurofibrillary tangle (NFT) (Holmes silver impregnation stain) Bar = 10 μm).

Several types of Aβ deposit have been identified in AD brains, but the majority can be classified into three morphological subtypes (Delaere et al., 1991; Armstrong, 1998): (1) diffuse deposits, in which most of the Aβ peptide is not aggregated into fibrils and dystrophic neurites and paired helical filaments (PHF) are infrequent or absent, (2) primitive deposits, in which the Aβ is aggregated into amyloid and dystrophic neurites and PHF are present, and (3) classic deposits, in which Aβ is highly aggregated to form a central 'core' surrounded by a 'ring' of dystrophic neurites. The latter type of deposit may be especially common in the visual cortex (Hof and Morrison, 1990; Armstrong et al., 1990).

The most important molecular constituent of the NFT is the microtubule associated protein (MAP) tau which is involved in the assembly and stabilization of the microtubules and therefore, establishes and maintains neuronal morphology (Lee et al., 1988; Roder et al., 1993). In normal neurons, tau is soluble and binds reversibly to microtubules with a rapid turnover (Caputo et al., 1992).

In AD, however, tau does not bind to the microtubules but collects as insoluble aggregates to form PHF which resist proteolysis and accumulate as NFT. Tau extracted from AD brains consists of both soluble and insoluble forms with, in the latter, the tau present in an abnormally phosphorylated isoform (Hanger et al., 1991).

VISUAL CHANGES IN ALZHEIMER'S DISEASE

A number of visual changes affecting many aspects of visual function have been reported in AD patients (Rumney, 1998) and have been reviewed by Katz and Rimmer (1989) and Fletcher (1994).

Visual Acuity

Visual acuity (VA) can be difficult to measure accurately in patients with AD during the later stages of the disease, but studies suggest that VA is normal in the early stages of the disease (Cogan, 1987; Sadun et al., 1987; Mendez et al., 1990; Rizzo et al., 1998). Low contrast acuity using Regan charts presented at four contrast levels, however, show a reduction in acuity with contrast in AD (Lakshminarayanan et al., 1996). In patients in whom impairments of VA have already been demonstrated, there may be impairments in stereopsis as measured by random dot stereograms (Kiyosawa et al., 1989).

Colour Vision

Some studies suggest colour vision is normal in patients with mild to moderate AD (Wood et al,., 1997). In other studies, however, defective colour vision may be present in approximately 50% of patients (Cogan, 1987; Mendez et al 1990). There is little available information on whether there are specific colour vision problems in AD, e.g., affecting the R/G rather than B/Y axis. However, the ability to use colour information may be impaired in AD. Hence, in cognitive tasks in which colour is used as an attention enhancer, a cue, or as a distracter, patients with AD were less accurate in their performance than controls (Wood et al., 1997).

Contrast Sensitivity

The contrast sensitivity function (CSF) provides a measure of performance across a wide range of spatial frequencies and contrasts. Some studies have not reported changes in contrast sensitivity in AD (Rizzo et al., 1998). Other studies, however, have reported reduced contrast sensitivities in

AD over all spatial frequencies (Lakshminarayanan et al., 1996, Nissen et al., 1985; Crow et al., 2003) and yet others reductions at lower spatial frequencies only (Ceccaldi, 1996). Differences in reported performance may be attributable to variation in patient population or assessment methods and especially the failure to account for VA differences between groups (Neargarder et al., 2003).

Critical Flicker Fusion Frequency Threshold and Visual Masking

The critical flicker fusion threshold, i.e., the lowest frequency at which the patient can no longer perceive a flickering light as a steady light, is normal in AD (Cronin-Golomb et al., 1991). Patients with AD often show deficits on visual masking tasks (Ceccaldi, 1996). Visual masking involves the presentation of a second stimulus immediately before (forward masking) or after (backward masking) a test stimulus. Patients with AD are significantly affected by a backward patterned mask stimulus compared with age-matched controls (Schlotterer et al., 1983). This result suggests that the speed of central visual processing is reduced in AD.

Visual Fields

There have been few studies of the visual fields in AD. In one report, visual sensitivity was reduced throughout the visual field but deficits were most pronounced in the inferior field (Trick et al., 1995). In addition, in follow up studies of patients with AD, there may be a progression of visual field loss over time (Trick et al., 1995).

Pupillary Function

Patients with AD may exhibit an abnormal pupillary response to the muscarinic receptor antagonist tropicamide, widely used by optometrists as a mydriatic (Scinto et al., 1994). Early reports suggested that some patients with AD display a specific response to low doses (typically 0.01%) of tropicamide, with pupils dilating by 13% or more compared with normal elderly controls. However, subsequent studies suggest that the tropicamide test is not a reliable method of diagnosing AD. There is no difference, for example, in the response of AD patients and those with vascular dementia (VaD) (Caputo et al., 1998) or Parkinson's disease (PD) (Granholm et al., 2003). In addition, not all studies have shown a significantly different response between AD and elderly control patients although a difference was demonstrated compared with young controls (Caputo et al., 1998). As a consequence, studies of the tropicamide response cannot be recommended as a clinical application to detect AD (Wilhelm et al., 1997; Graff-Radford et al., 1997).

Enhanced pupillary responses in AD have also been reported in response to the application of dilute solutions of phenylephrine (a sympathetic agonist) and to pilocarpine (a cholinergic agonist) (Hanyu et al., 2007). Reductions in the pupillary light reflex have also been reported in AD (Tales et al., 2001).

Eye Movements

The ability of some patients to fixate a target is affected in AD (Sadun et al., 1987; Mendez et al., 1990). Defects of fixation control may be associated with degeneration of the parietal lobe, a region believed to be involved in maintaining fixation stability. Several changes in saccadic eye movements have been reported. First, saccadic latency declines with age but the delays become more pronounced in AD (Fletcher and Sharpe, 1988). Second, saccadic velocities are reduced, the degree of this reduction being correlated with the severity of the dementia. Third, patients at more advanced stages of the disease exhibit inaccurate saccades with undershooting of the target by 10-30% (Sadun and Bassi, 1990). Fourth, 50% of patients with AD will have difficulty in initiating or maintaining saccadic eye movements (Sadun and Bassi, 1990).

Smooth pursuit eye movements, which are a sensitive indicator of brain function, may also be affected (Zaccara et al., 1992). A gradual deterioration of these movements occurs in AD with catch up saccades being necessary to maintain fixation (Fletcher and Sharpe, 1988). Degeneration and atrophy of the frontal and/or the parietal lobes may be responsible for these changes.

Electrophysiology

The amplitude of the pattern electroretinogram (PERG) response is reduced in AD although the flash ERG may be unaffected (Katz and Rimmer,

1989; Trick et al., 1989). A significant delay in the N35, P50, and N95 components of the PERG together with reductions in amplitude have been reported accompanied by significant reductions in nerve fibre layer thickness (Parisi et al., 2001). In other studies the amplitude and latency of components of the PERG have been reported to be unaffected (Kergoat et al., 2002; Strenn et al., 1991; Prager et al., 1993). The reduction in the 'b' wave shown in some studies could be attributable to the reduced number of ganglion cells in the retina in AD (Parisi et al., 2001; Hinton et al., 1986).

Second, a number of studies dating from the 1980s have suggested that the latency of the flash P2 component of the cortical visual evoked response (VEP) is delayed and the P100 component to a reversing checkerboard is normal in patients with AD (Philpot et al., 1990; Harding et al., 1985; O'Neil et al., 1989). This pattern of abnormalities, however, has not been shown in all studies and has not been adopted as a routine test for AD (Coben et al., 1983).

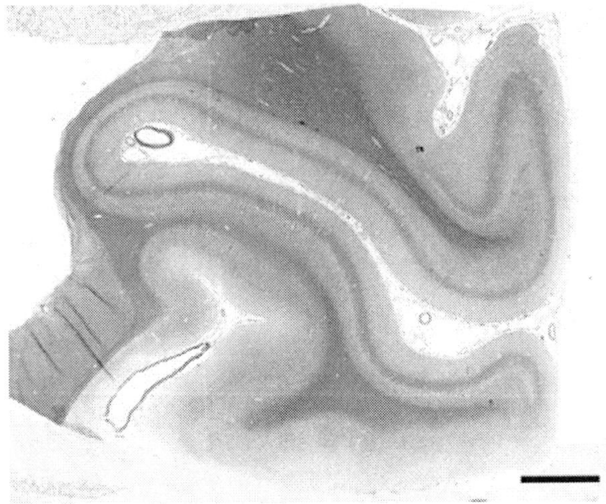

Figure 2. Coronal section of the occipital cortex showing the Band of Gennari which distinguishes the primary visual area V1 from V2. (Holmes silver impregnation stain, Magnification bar = 2mm).

Complex Visual Functions

Patients with AD often exhibit difficulties with reading (Glosser et al., 2002), visuospatial function (Fujimori et al., 1997; Geldmacher, 2003) and in the identification and naming of objects (Cogan, 1987). Significantly greater thresholds for perceiving shapes defined by motion cues (Ceccaldi, 1996) may be present and this is likely to affect object recognition. Patients with AD also show impairment of eye-head coordination (Nakano et al., 1999), problems with finding objects when surrounded by others (Tales et al., 2002), and in finding known objects in an unknown environment (Nguyen et al., 2003). Deficient perception and cognition in AD are often attributed to slow information processing. Increasing 'stimulus strength' by increasing stimulus contrast can often improve various aspects of the cognitive performance in AD (Gilmore et al., 2005).

In rare cases, patients develop a combination of visual symptoms called 'Balint's syndrome' (Chapman et al., 1999). These include a psychic paralysis of gaze (ocular apraxia), optic ataxia (lack of muscular coordination) e.g., an inability to guide the hand towards an object using visual information, and a spatial disorder of attention ('simultanagnosia'), viz., an inability to report all items or their relationships in pictures depicting events or situations (Fletcher, 1994). These symptoms are accompanied by visual field constriction, the fading of centrally fixated objects, and impaired reading ability despite normal VA. It is possible that these symptoms are attributable to degeneration of more complex areas of the visual cortex (Fletcher, 1994).

Visual Hallucinations

Visual hallucinations may be present in some patients and especially those with impaired VA and with more severe cognitive impairment (Chapman et al., 1999).

The Visual Cortex in AD

Pathological changes within the visual cortex in AD normally involve the visual association areas and spare to some degree, the primary visual cortex (area V1). However, a number of pathological changes have been reported in the visual cortex in Alzheimer patients. In a retrospective study of 106 patients with AD, SP and NFT were observed in the visual cortex in 72% and 27% of cases respectively. The density of SP and especially NFT is generally greater in the visual association areas (V2, V3 etc.) than in V1 especially in younger cases (Armstrong et al., 1990). In area V1, SP with distinct amyloid cores and relatively few NFT can be observed while in V2, numerous uncored 'neuritic' plaques (NP) and NFT are usually present (Hof and Morrison, 1990). Whether differences in cortical pathology between areas V1 and V2 are responsible for the cortical VEP to flash and pattern stimuli described previously remains to be established (Armstrong, 1994). In a significant proportion of cases of AD, the density of SP and/or NFT is significantly greater in the cuneal compared with the lingual gyrus of area V1 (Armstrong, 1996). This difference could contribute to the predominantly inferior visual field deficits reported by Trick et al. (1995). In addition, the amount of myelin appears to be reduced in the outer laminae of the visual cortex and loss of neurons and neurotransmitters have also been reported (Hof and Morrison, 1990).

Objectives

In many neurodegenerative disorders, the density of pathological lesions varies across the cortical laminae from pia mater to white matter. The laminar distribution of a lesion may reflect degeneration of neural pathways that have their cells of origin or axon terminals located within a particular lamina (De Lacoste and White 1993, Armstrong and Slaven, 1994). Hence, to characterize further the pattern of degeneration of the visual cortex in AD, the distribution of the SP and NFT across the cortex from pia mater to white mater was studied in 18 cases of AD using quantitative methods. The specific objectives were: (1) to describe the changes in density of SP and NFT across the laminae from pia mater to white matter, (2) to determine whether the laminar distribution of the SP and NFT was correlated with patient age, disease onset or disease duration, (3) to determine whether SP and NFT may spread between visual cortical areas via the cortico-cortical pathways, and (4) to determine which aspects of visual processing are likely to be affected in AD.

Materials and Methods

Cases

Tissue from 18 cases (details in Table 1) of AD was obtained from the Dept of Neuropathology, University of Washington, Seattle. The principles embodied in the Helsinki declaration were followed with respect to experiments involving material of human origin. Patients were clinically assessed and all fulfilled the 'National Institute of Neurological and

Communicative Disorders and Stroke and the Alzheimer Disease and Related Disorders Association' (NINCDS/ADRDA) criteria for probable AD (Tierney et al., 1988). The histological diagnosis of AD was established by the presence of widespread neocortical SP consistent with the 'Consortium to Establish a Registry of Alzheimer Disease' (CERAD) criteria (Mirra et al., 1991) and 'National Institute on Aging (NIA)-Reagan Institute' criteria (Jellinger and Bancher, 1998).

Table 1. Demographic and general pathological features of the Alzheimer's disease (AD) cases studied

Case	Age	Onset	Duration	Gender	PM	Cause of death
A B	65 57	60 52	5 3	M F	13 2	Myocardial infarction Metastatic carcinoma
C D	88 82	58 NA	30 NA	F M	NA NA	Bronchopneumonia Colon carcinoma
E F	61 78	NA 73	NA 5	M M	NA 21	Bronchopneumonia Chronic bronchial disease
G	85	75	10	M	3	Stroke
H	80	65	15	M	9	Stoke
I	83	78	5	M	54	Stroke
J K	84 79	65 NA	19 NA	M F	3 NA	Bronchopneumonia Stroke
L	86	NA	NA	M	NA	NA
M	69	49	20	F	NA	NA
N O	72 79	52 71	20 8	M M	NA 48	Bronchopneumonia Stroke
P Q R	63 72 77	56 65 67	7 7 10	M F F	NA NA 4	Lymphatic leukaemia Pulmonary embolism Coronary artery disease

Abbreviations: M = Male, F = female, NA = data not available, PM = post-mortem delay time (hours).

Histological Methods

A block of tissue (approximately 20 x 25 mm in the coronal plane, 20 mm thick) was removed from the medial surface of the brain from each case approximately half way along the length of the calcarine fissure. The block included the calcarine fissure (area V1), the cuneal and lingual gyri, and at least one adjacent gyrus (area V2) on each side. Tissue was fixed in 10% formal-saline for 2-3 weeks and 10 μm sections were cut and stained by the Holmes silver impregnation method to reveal the SP and NFT. Post-mortem delay varied from under two hours to 54 hours (Table 1) but there was no evidence that variations in post-mortem delay affected the quality of silver staining. Adjacent sections were stained with cresyl violet to enable cortical laminae to be clearly identified.

Identification of Visual Areas

Identification of visual areas was based on Brodmann's cortical map. Area V1 has well-defined boundaries although the detailed architecture is not uniform throughout its length. V1 is typical of granular cortex in which the laminar pattern is partially obscured because of the high density of granule cells present. It can be identified by the presence of the Band of Gennari (Figure 2) and by the cell poor sublamina IVB (Clarke and Miklossy, 1990). Area V2 occupies a 'horse-shoe' shaped region surrounding V1 and has a clearer six-layered structure in which layer III is broader than IV, the latter with a high density of small pyramidal cells (Clarke and Miklossy, 1990). In addition, V2 can be identified by the large pyramidal cells in the deeper parts of lamina III (Clarke and Miklossy, 1990). The boundary between V2 and V3 is more difficult to detect as microscopic changes occur more gradually. Hence to delimitate area V2 from V3, the position of V2 was determined approximately by comparison of the morphology of the occipital cortex with Brodmann's cortical maps (Clarke and Miklossy, 1990).

Quantifying the Pathological Features

The laminar distribution of the pathological changes across the cortex was studied in areas V1 and V2 of each AD case using methods based on those of Duyckaerts et al. (1986). Five traverses from the pia mater to white matter were located at random within each visual area. The density of SP and NFT was recorded at a magnification of x400 in 250 x 125 μm sample fields, the larger dimension of the field being located parallel with the surface of the pia mater. The eye piece micrometer was moved down each traverse one step at a time using histological features to correctly position the field. Depending on the degree of cortical atrophy, between 10 and 14 sample fields were necessary to sample each traverse. Counts from the five traverses were then added together to study the laminar distribution of lesions within each region. The boundary between layers I and II and the location of laminae IV were determined from the cresyl violet sections. The location of the II/III and V/VI boundaries were more difficult to establish from the sections. Hence, the approximate position of these boundaries was established based on the laminar profiles of V1 and V2 in Brodal (1981) and Clarke and Miklosy (1990).

Statistical Analysis

The data comprise density measurements of the densities of SP and NFT in each cortical lamina 1 to 6 (I to VI) in areas V1 and V2 of the visual cortex in 18 cases of AD. The SP and NFT data were analysed separately by two-factor, split-plot analysis of variance (ANOVA) with visual area as the main-plot factor and cortical lamina as the sub-plot factor (Snedecor and Cochran, 1980). In addition, the relationship between the density of SP and NFT in each lamina and patient age, disease onset, and disease duration was tested by Pearson's correlation coefficient ('r'). Correlations between densities of SP and NFT in laminae of V1 and V2 were also examined.

Laminar Distribution of the SP and NFT in V1 and V2

Variation in density of SP and NFT across areas V1 and V2 in a single case of AD are shown in Figure 3. In area V1, low densities of SP were present with SP distributed across the cortex with slightly higher densities in lamina III while the distribution of the NFT was bimodal with a large density peak in lamina III and a smaller density peak in lamina IV. In area V2, SP were restricted to the upper laminae while the distribution of NFT was bimodal with a large density peak in lamina II and a smaller density peak in lamina V.

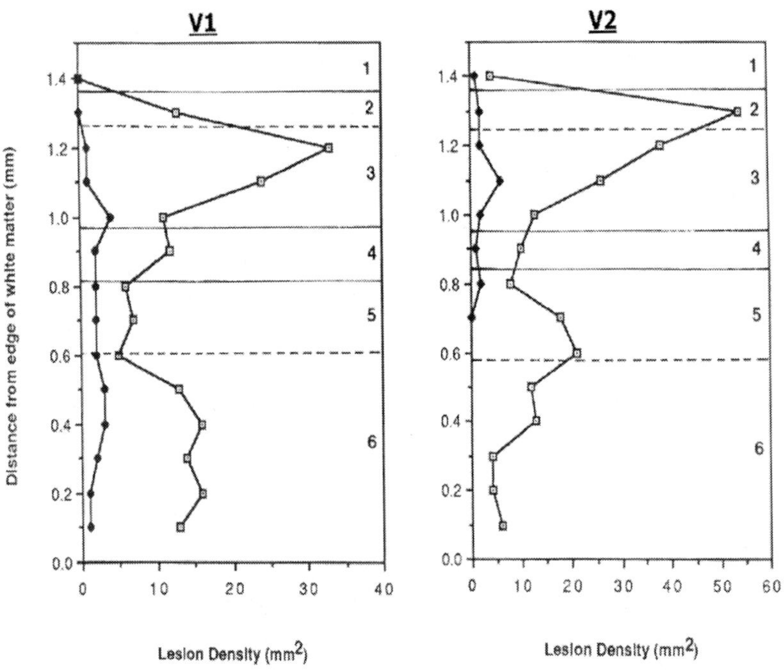

Figure 3. Distribution of senile plaques (SP) (dark symbols) and neurofibrillary tangles (NFT) (open symbols) across the cortex from pia mater to white matter in areas V1 and V2 in a case of Alzheimer's disease (AD). Horizontal bands indicate the approximate locations of the different cortical laminae (1 – 6).

The mean densities of SP in each cortical lamina in areas V1 and V2 in each of the 18 AD cases together with their overall mean densities are shown in Tables 2 and 3. The ANOVA suggested there were no consistent differences in the density of SP between areas V1 and V2 ($F = 3.31$, $P > 0.05$). However, significant differences in density were observed between cortical laminae ($F = 16.88$, $P < 0.001$), the non-significant interaction ($F = 3.19$, $P > 0.05$) indicating that laminar distributions were essentially similar in V1 and V2. In V1, SP density was similar in laminae II, IV, V, and VI while lamina I appeared to have the lowest SP density. A similar distribution of SP was observed in area V2, except that SP density in lamina IV was not significantly different to that in lamina III.

Table 2. The mean density (per 250 x 125 μm field) of senile plaques (SP) in each cortical lamina in area V1 of the visual cortex in Alzheimer's disease (* indicates peak density where a distinct peak is evident)

Laminae of area V1

Case	Lesion	I	II	III	IV	V	VI
A	SP	0	0.4	1.0*	0.3	0.4	0.2
	NFT	0	0	0.1	0	0.1	0.1
B	SP	0.4	1.6	1.7*	0.8	0	0.7
	NFT	0	0.4	0.3	0.4	0.3	0
C	SP	0	0	0.1	0.2	0.5*	0.4
	NFT	0	0	0	0.2	0.3	0.1
D	SP	0	0	0.2	0.6*	0.6*	0.4
	NFT	0	0	0.2	0.2	0.2	0
E	SP	0	0	0.4	0.9*	0.6	0.1
	NFT	0	0	0.4	0.6	0.8	1.2*
F	SP	0	0	0	0	0	0.1
	NFT	0.2	0	0	0	0	0.1
G	SP	0	0	0.4	0.6*	0.6*	0.3
	NFT	0	0	0.2	0.2	0	0.3
H	SP	0	0.2	0.7*	0.4	0	0
	NFT	0	0	0	0	0	0
I	SP	0	0	0.7*	0	0	0
	NFT	0	0.2	0.1	0	0	0
J	SP	0	0.8	1.4*	0.7	0.3	0
	NFT	0	2.0*	0.8	0	0	0.2
K	SP	0	1.0*	0.8	0.2	0.4	0.1
	NFT	0	0	0.6*	0.2	0.1	0.1
L	SP	0	0	0	0.4	0.2	0
	NFT	0	0	0	0	0.1	0
M	SP	0	0.2	0.9*	0.2	0.6	0.1
	NFT	0	2.0	2.5*	0.6	0.1	0
N	SP	0	0.2	0.4	0.3	0.2	0
	NFT	0	0.2	0.5*	0	0	0.2
O	SP	0	0	0	0.2	0	0

Table 2. (Continued)

Case	Lesion	I	II	III	IV	V	VI
	NFT	0	0	0.4	0.6	0.1	0.7*
P	SP	0	0.4	1.3*	0.4	0.8	0.2
	NFT	0	4.4*	1.9	1.8	3.5	1.4
Q	SP	0	0	0.4	0.4	0.4	0.4
	NFT	0	2.6	4.6*	1.9	1.3	2.8
R	SP	0	0.8	2.3*	0.6	0	0.4
	NFT	0	0.8	0.4	0.4	0	0.1
Mean	SP	0.02	0.31	0.71	0.40	0.31	0.19
	NFT	0.01	0.70	0.72	0.39	0.38	0.41

Table 3. The mean density (per 250 x 125 μm field) of neurofibrillary tangles (NFT) in each cortical lamina in area V2 of the visual cortex in Alzheimer's disease (*indicates peak density where a distinct peak is evident)

Laminae of area V2

Case	Lesion	I	II	III	IV	V	VI
A	SP	0.3	0.3	0.9*	0.6	0	0.2
	NFT	0	2.0*	0.9	0.1	0.2	0
B	SP	0.2	1.4*	1.1	0.7	1.0	0.5
	NFT	0.2	4.4	4.9*	1.7	2.0	0.9
C	SP	0	0	0.9*	0.6	0.5	0.4
	NFT	0	0.8*	0.2	0	0.3	0.3
D	SP	0	0	0.6	0.6	0.7*	0.3
	NFT	0	0.2	0.1	0.2	0.1	0.2
E	SP	0	0	0.6	0.6	0.7*	0.3
	NFT	0	0.2	1.1*	1.0	0.9	1.1
F	SP	0	0.8*	0.5	0.4	0	0
	NFT	0	0	0.8*	0.2	0.2	0
G	SP	0	0	0.3	0.2	0.2	0.5
	NFT	0	0.4*	0	0	0	0.1
H	SP	0.2	0.2	0.8*	0.2	0.2	0.3

Case	Lesion	I	II	III	IV	V	VI
	NFT	0	0	0	0	0	0
I	SP	0.2	1.0*	0.9	0.8	0	0
	NFT	0.2	0.4	0.1	0.3	0	0
J	SP	0	0	1.1	1.5*	1.0	0.6
	NFT	0	0.6	3.9*	2.4	0.5	0.7
	NFT	0	0.4	0.3	0.4	0.4	0
L	SP	0	0	0.1	0.6*	0.8*	0.3
	NFT	0.4	0.4	0	0.2	0.4	0
M	SP	0	0	0.5	0.9	1.2*	0.2
	NFT	0.4	0.4	1.8	0.8	2.2*	0.4
N	SP	0.6	0.8*	0.4	0.6	0	0.2
	NFT	0	1.8	4.7*	1.6	1.2	0.7
O	SP	0.2	0	0	0	0	0
	NFT	0	0	0.2	0	0.2	0.2
P	SP	0	1.0	2.7*	2.2	0.8	0.6
	NFT	0	1.0	1.3*	0.8	0.2	1.2
Q	SP	0.2	0.4	0.7*	0.4	0	0
	NFT	0.8	11.0*	5.1	2.2	3.8	1.5
R	SP	0	0.7*	0.5	0	0	0
	NFT	0.2	3.4	2.8	1.0	1.6	0.2
Mean	SP	0.10	0.39	0.75	0.65	0.41	0.25
	NFT	0.12	1.52	1.57	0.72	0.79	0.42

The ANOVA suggested that the density of NFT was greater in area V2 than V1 ($F = 4.46$, $P < 0.05$). In addition, significant variations in density were observed between cortical laminae ($F = 7.86$, $P < 0.001$) and these differences were similar in V1 and V2 ($F = 1.40$, $P < 0.05$). In V1, the density of NFT was greatest in laminae II and III and least in lamina I while laminae IV, V, and VI had a similar NFT density. The distribution of NFT was similar in B18 except that the density in laminar VI was significantly less than in IV and V.

The data suggested that the density of NFT was greater in V2 than V1 whereas the density of SP was similar in both regions as reported previously (Rogers and Morrison, 1985; Lewis et al., 1987; Hof et al 1989, Braak et al, 1989). In V1, mean densities of SP and NFT were maximal in laminae III and laminae II and III respectively. These data are in agreement with those of Hof and Morrison (1990) who also found that NFT in V1 were located predominantly in laminae II and III. In addition, the data agree in that the majority of SP were supragranular but disagree in suggesting that SP were predominantly located in lamina III rather than lamina II. By contrast, Beach and McGeer (1988) found that in area V1, SP were aggregated at the interface of layers IVc and V. In area V2, mean density of SP was maximal in III and IV and NFT in laminae II and III. By contrast previous authors have found that SP were predominant in laminae II and III (Hof and Morrison, 1990) while NFT were most frequently observed in lamina III (Rogers and Morrison, 1985; Lewis et al., 1987; Hof et al 1989, Braak et al, 1989). These differences could be attributable to differences in the patient population or staining procedures used.

Correlations with Age, Disease Onset and Duration

The correlations (Pearson's 'r') between the densities of SP and NFT in each lamina of areas V1 and V2 and patient age, disease onset, and duration are shown in Table 4. The densities of SP in laminae I of V1 and NFT in lamina IV of V2 were negatively correlated with patient age. No significant correlations were observed in any cortical lamina between the density of NFT and disease onset or duration. However, in V2, the density of SP in lamina II and lamina V was negatively correlated with disease duration and disease onset respectively.

The pattern of correlations with patient age and duration in these patients throws little light on the development of laminar pathology in the visual cortex in AD. However, the negative correlations between the density of SP and NFT and patient age and duration of disease suggest that some lesions could be lost as the disease progresses.

Armstrong et al. (1990) found evidence that the overall density of SP and NFT declined with age in several cortical areas in AD including V1 and V2 consistent with these data. In addition, lesions could spread vertically within the columns of the visual cortex and then disappear within some laminae as the disease progresses (Johnson and Blum, 1970; Probst et al 1982). Hence, the original laminar distribution of SP and NFT could be obscured by later developments especially in older and long duration cases.

Table 4. Correlations (Pearson's 'r') between the densities of senile plaques (SP) and neurofibrillary tangles (NFT) in areas V1 and V2 of the visual cortex in Alzheimer's disease (AD) and patient age, disease onset, and disease duration (* P < 0.05)

		SP			NFT		
Region	Lamina	Age	Onset	Duration	Age	Onset	Duration
V1	I	-0.50*	-0.36	-0.25	0.07	0.31	-0.25
	II	-0.40	-0.42	-0.16	-0.34	-0.34	-0.05
	III	-0.40	-0.27	-0.18	-0.30	-0.27	-0.06
	IV	-0.34	-0.24	-0.02	-0.45	-0.21	-0.24
	V	-0.17	-0.37	0.28	-0.45	-0.26	-0.21
	VI	-0.30	-0.27	-0.14	-0.33	0.02	-0.26
V2	I	-0.26	-0.24	-0.06	-0.09	-0.11	-0.16
	II	-0.47	-0.12	-0.52*	-0.32	-0.15	-0.25
	III	-0.39	-0.32	-0.11	-0.46	-0.47	0.01
	IV	-0.27	-0.38	0.07	-0.38	-0.36	0.05
	V	-0.23	-0.60*	0.34	-0.43	-0.40	-0.03
	VI	-0.08	-0.38	0.30	-0.60* -	0.47	-0.02

Correlations between SP and NFT in Areas V1 and V2

The significant correlations (Pearson's 'r') between the densities of SP in areas V1 and V2 with those of NFT in areas V1 and V2 is shown in Table 5. Several correlations between the different visual areas were apparent including: (1) a positive correlation between the density of SP in lamina I of area V1 with the density of NFT in lamina II of area V2, (2) a positive correlation between the density of SP in lamina II of area V1 with the density of NFT in lamina III of area V2, 3) positive correlations between the densities of SP in lamina III of V2 with NFT in laminae II, IV, and V of V1, and (4) positive correlations between the density of SP in lamina IV of area V2 and NFT in laminae II and V of area V1.

These data provide some support for the hypothesis that the formation of SP and NFT in AD are interconnected, i.e., that SP may develop at the axon terminals of NFT-containing cells (Armstrong et al., 1993).

Table 5. Significant correlations (Pearson's 'r') between the densities of senile plaques (SP) and neurofibrillary tangles (NFT) between areas V1 and V2 of the visual cortex in Alzheimer's disease (AD) (* P < 0.05, **P < 0.01, ***P < 0.001)

Area and lesion correlated		Significant correlations
Area V1	Area V2	
SP	NFT	(1) SP (lamina I) and NFT (lamina II) r = 0.53*
		(2) SP (lamina II) and NFT (lamina III) r = 0.48*
NFT	SP	(1) SP (lamina III) and NFT (lamina II) r = 0.69**
		(2) SP (lamina III) and NFT (lamina IV) r = 0.47*
		(3) SP (lamina III) and NFT (lamina V) r = 0.79***
		(4) SP (lamina IV) and NFT (lamina II) r = 0.72***
		(5) SP (lamina IV) and NFT (lamina V) r = 0.67**

A high density of NFT has been observed in subcortical nuclei projecting to SP-rich areas in the neocortex (Mann, 1985), while a correlation has been found between SP in the molecular layer of the dentate gyrus and NFT in cells of lamina II of the entorhinal cortex (EC) consistent with this hypothesis. However, Armstrong (1992) found only weak correlations between the density of cellular NFT in various regions of the hippocampus and the sum of the β-amyloid (Aβ) deposits at their target sites.

Does the Pathology Spread between Visual Cortical Areas?

De Lacoste and White (1993) have proposed two hypotheses to explain the possible spread of SP and NFT from an origin in the medial temporal lobe in AD. First, that the degeneration initially involves axonal terminals in the entorhinal cortex (EC) and spreads retrogradely towards cell bodies in the rest of the parahippocampal gyrus (PHG).

The degeneration then spreads to affect other regions of the temporal lobe and ulitmately the visual areas V2 and V1 (feedforward sequence). Second, the degeneration initially involves cell bodies in the EC, spreads along an orthograde route to the rest of the PHG and then to the primary sensory areas (feedback sequence). Since cortico-cortical connections arise and terminate in predictable cortical laminae, it should be possible to test both hypotheses using the present data. The data are more consistent with the spread of SP and NFT between V2 and V1 via the feedforward rather than the feedback connections. If the degeneration proceeds retrogradely in relation to the feedforward projections, NFT should be found in the supragranular laminae of areas V1 and V2 while most SP should be in lamina IV in area V2 (De Lacoste and White, 1993). The data suggest a NFT peak in the supragranular laminae in areas V1 and V2 as predicted by the hypothesis. In addition, SP density in V2 was maximal in laminae III and IV. The high density of SP in lamina IV is consistent with the hypothesis. The SP in lamina III in area V2 may reflect their formation: (1) on the dendritic trees of lamina IV neurons which project into lamina III, (2) on the dendritic trees of lamina III neurons as a result of the formation of NFT, or (3) on the axon terminals of subcortical NFT-containing cells (Mann, 1985). If the degeneration occurs in the orthograde direction in relation to the feedback projections, then the hypothesis predicts that NFT should be distributed mainly in the infragranular laminae within area V1, SP being present in all laminae but especially in laminae I (De Lacoste and White, 1993) In area V2, however, maximum NFT density occurred in laminae II and III. In addition, there were no cases with a SP peak of density in lamina I, maximum SP density often occurring in lamina III.

Neurodegeneration in AD could also occur in conjunction with the longer cortico-cortical and the cortical-subcortical connections (Hof and Morrison, 1990). In area V1, the long cortico-cortical projections originate from cells in lamina IVb, the adjacent region of lamina III and from laminae V/VI and in area V2, from laminae III and V (Hof and Morrison, 1990). Hence, the development of some cellular NFT at these sites may be related to the formation of SP at more distant projection sites.

Which Aspects of Visual Processing Are Likely to Be Affected in AD?

Visual information arrives in the striate cortex from the lateral geniculate nucleus (LGN) and terminates in lamina IVa and IVb (Singer, 1979) (Figure 4), then being conveyed to laminae II and III and to a lesser extent lamina V.

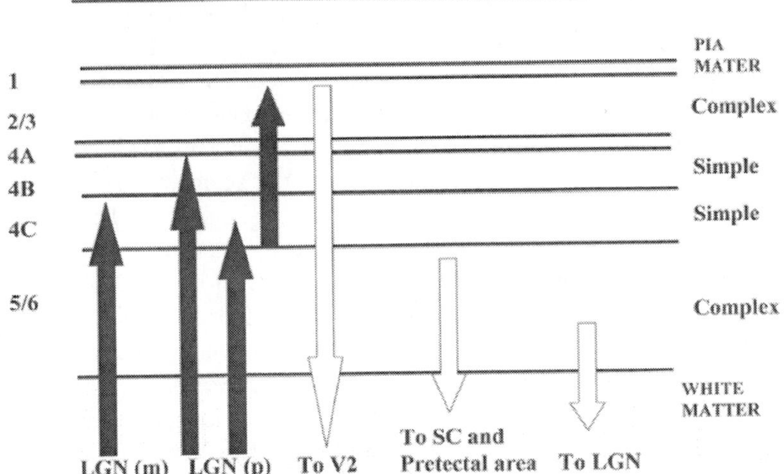

Figure 4. Laminar structure of the visual cortex (area V1) showing the afferent and efferent pathways. (LGN = Lateral geniculate body, SC = Superior colliculus). Lamina IV cells respond to relatively simple visual stimuli whereas cells in laminae II/III and V/VI have more complex responses.

The output from area V1 to V2 is mainly via the large pyramidal neurons in laminae II/III which constitute the feedforward cortico-cortical projections (De Lacoste and White, 1993).

In addition, there is a connection from area V1 originating in lamina IVb to the medial temporal area (MT or V5) which is crucial for the analysis of motion in the visual field (Zeki and Shipp, 1988). Area V5 also receives a projection from the deep part of lamina III originating in V2 (Shipp and Zeki, 1989). In area V1, there was a significant laminar degeneration associated with the development of SP in lamina III and NFT in laminae II and III. These pathological changes are likely to affect visual processing within the upper laminae in area V1 and subsequently, the transfer of information to area V2. In addition, the pathological changes in the lower laminae could affect the neurons providing the output to subcortical visual areas, e.g., from lamina VI to the dorsal and ventral LGN and from lamina V to the superior colliculus and the pretectal area (Parnavelas and McDonald, 1983). A similar distribution of the pathological changes is present in area V2 with the exception that the density of SP was maximal in laminae III and IV. These results suggest that the short cortico-cortical pathways connecting areas V1 and V2 and the output from these areas to subcortical visual areas are likely to be compromised in

AD. By contrast, the long cortico-cortical pathway from area V1 to V5 is less likely to be affected although there is some evidence that the projection from area V2 to V5 may be compromised.

Conclusion

There is a distinct pattern of laminar pathology in the visual cortex in AD. In area V1, the mean density of SP and NFT reached a maximum in lamina III and in laminae II and III respectively. In area V2, mean SP density was maximal in laminae III and IV and NFT density in laminae II and III. The densities of SP in laminae I of area V1 and NFT in lamina IV of area V2 were negatively correlated with patient age. No significant correlations were observed in any cortical lamina between the density of NFT and disease onset or duration. However, in area V2, the density of SP in lamina II and lamina V was negatively correlated with disease duration and disease onset respectively. In addition, there were several positive correlations between the densities of SP and NFT in area V1 with those in area V2. The data suggest that the pathology may spread between visual areas via the feed-forward short cortico-cortical connections. The longer cortico-cortical and cortico-subcortical connections may also be affected by the pathology of AD. Pathological changes in different regions of the visual cortex may contribute to several of the visual problems identified in patients with AD.

Acknowledgements

The assistance of the Department of Neuropathology, University of Washington, Seattle, USA in providing cases for this study is gratefully acknowledged.

References

Alzheimer A. On a peculiar disease of the cerebral cortex. *Allgemeine Zeitschrift fur Psychiatrie und Psychish-Gerichtlich Medicin* 1907, 64, 146-148.

Armstrong RA. Alzheimer's disease: Are cellular neurofibrillary tangles linked to β/A4 formation at the projection sites? *Neurosci. Res. Commun.* 1992, 11, 171-178.

Armstrong RA. Neuropathological differences between areas B17 and B18: Implications for visual evoked responses in Alzheimer's disease. *Dementia* 1994, 5, 247-251.

Armstrong RA. Visual field defects in Alzheimer's disease patients may reflect differential pathology in the primary visual cortex. *Optometry and Vision Science,* 1996, 73, 677-682.

Armstrong RA. β-amyloid plaques: stages in life history or independent origin? *Dement and Ger Cog Disord* 1998, 9, 227-238.

Armstrong RA. The interface between Alzheimer's disease, normal aging, and related disorders. *Curr Aging Sci* 2008, 1, 122-132.

Armstrong RA, Nochlin D, Sumi SM, Alvord EC. Neuropathological changes in the visual cortex in Alzheimer's disease. *Neuroscience Research Communications,* 1990, 6, 163-171.

Armstrong RA, Myers D, Smith CUM. The spatial pattern of plaques and tangles in Alzheimer's disease do not support 'the cascade hypothesis'. *Dementia* 1993, 4, 16-20.

Armstrong RA, Slaven A. Does the neurodegeneration of Alzheimer's disease spread between visual cortical regions B17 and B18 via the feedforward or feedback short cortico-cortical projections ? *Neurodegen* 1994, 3, 191-196.

Beach TG, McGeer EG. Lamina-specific arrangement of astrocytic gliosis and senile plaques in Alzheimer's disease. *Brain Res.* 1988, 463, 357-361.

Braak H, Braak E, Kalus P. Alzheimer's disease : areal and laminar pathology in the occipital isocortex. *Acta Neuropathol.* 1988, 77, 494-506.

Brodal A, 1981. Neurological Anatomy. Oxford University Press, Oxford.

Brun A. An overview of light and electron microscope changes. In: Reisberg B, editor. *Alzheimer's disease: The Standard Reference.* London and New York: MacMillan, 1983, 37-45.

Caputo CB, Sygowski LA, Scott CW, Evangelista-Sobel IR. Role of tau in the polymerization of peptides from β-amyloid precursor protein. *Brain Res* 1992, 597, 227-232.

Caputo L, Casartelli M, Perrone C, Santori M, Annoni G, Vergaini C. The 'eye-test' in recognition of late-onset Alzheimer's disease. *Arch. Geront Ger* 1998, 27, 171-177.

Ceccaldi M. Vision in Alzheimer's disease. *Revue Neurol.* 1996, 152, 6-7.

Chapman FM, Dickinson J, McKeith I, Ballard C. Association among visual associations, visual acuity, and specific eye pathologies in Alzheimer's disease: Treatment implications. *Am. J. Psych.* 1999, 156, 1983-1985.

Clarke S, Miklossy J. Occipital cortex in man: organization of callosal connections, related myelo- and cytoarchitecture and putative boundaries of functional visual areas. *Journal of Comparative Neurology,* 1990, 298, 188-214.

Coben LA, Danziger WL, Hughes CP. Visual evoked potentials in mild senile dementia of Alzheimer type. *Electroenceph. Clin. Neurophysiol* 1983, 55, 121-130.

Cogan DG. Alzheimer syndromes. *Am. J. Ophthalmol.* 1987, 104, 183-184.

Cronin-Golomb A, Corkin S, Rizzo JF, Cohen J, Growden JH Banks KS. Visual dysfunction in Alzheimer's disease: relation to normal ageing. *Ann. Neurol.* 1991, 29, 41-52.

Crow RW, Levin LB, LaBree L, Rubin R, Feldon SE. Sweep visual evoked potential evaluation of contrast sensitivity in Alzheimer's dementia. *Inv. Opthalmol. Vis. Sci.* 2003, 44, 875-878.

De Lacoste M, White CL III. The role of cortical connectivity in Alzheimer's disease pathogenesis: a review and model system. *Neurobiology of Aging,* 1993, 14, 1-16.

Delaere P, Duyckaerts C, He Y, Piette F, Hauw J. Subtypes and differential laminar distributions of β/A4 deposits in Alzheimer's disease: relationship with the intellectual status of 26 cases. *Acta Neuropathol* 1991, 81, 328-335.

Duyckaerts C, Hauw JJ, Bastenaire F, Piette F, Poulain C, Rainsard V, Javoy-Agid F, Berthaux P. Laminar distribution of neocortical senile plaques in senile dementia of the Alzheimer type. *Acta Neuropathol,* 1986, 70, 249-256.

Evans DA, Funkenstein H, Abert MS et al. Prevalence of Alzheimer's disease in a community of older persons - higher than previously reported. *J Am Med Assoc* 1989, 262, 2551-2556.

Ferri CP, Prince M, Brayne C, Brodaty H, Fratiglioni L, Ganguly M et al. Global prevalence of dementia: a Delphi consensus study. *Lancet* 2005, 366, 2112-2117.

Fletcher WA. Ophthalmological aspects of Alzheimer's disease. *Curr. Opin. Ophthalmol.* 1994, 5, 38-44.

Fletcher WA, Sharpe JA. Smooth pursuit dysfunction in Alzheimer's disease. *Neurology* 1988, 38, 272-277.

Fujimori M, Imamura T, Yamashita H, Hirono N, Mori E. The disturbances of object vision and spatial vision in Alzheimer's disease. *Dement and Ger Cog. Disord.* 1997, 8, 228-231.

Geldmacher DS. Visuospatial dysfunction in the neurodegenerative diseases. *Frontiers in Bioscience* 2003, 8, E428-E436.

Gilmore GC, Cronin-Golomb A, Neargarder SA, Morrison SR. Enhanced stimulus contrast normalizes visual processing of rapidly presented letters in Alzheimer's disease. *Vis. Res.* 2005, 45, 1013-1020.

Glenner GG, Wong CW. Alzheimer's disease and Down's syndrome: sharing of a unique cerebrovascular amyloid fibril protein. *Biochem. Biophys. Res. Commun* 1984, 122, 1131-1135.

Glosser G, Baker KM, de Vries JJ, Alavi A, Grossman M, Clark CM. Disturbed visual processing contributes to impaired reading in Alzheimer's disease. *Neuropsychologia* 2002, 40, 902-909.

Graff-Radford NR, Lin SC, Brazis PW, Bolling JP, Liesegang TJ, Lucas JA et al. Tropicamide eyedrops cannot be used for reliable diagnosis of Alzheimer's disease. *Mayo Clin. Proc.* 1997, 72, 495-504.

Granholm E, Morris S, Galasko D, Shults C, Rogers E, Vukov B. Tropicamide effects on pupil size and papillary light reflexes in Alzheimer's and Parkinson's disease. *Int. J. Psychophysiol.* 2003, 47, 95-115.

Hanger DP, Brion JP, Gallo JM, Cairns NJ, Luthert PJ, Anderton BH. Tau in Alzheimer's disease and Down's syndrome is insoluble and abnormally phosphorylated. *Biochem J.* 1991, 275, 99-104.

Hanyu H, Hirao K, Shimizu S, Kanetaka H, Sakurai H, Iwamoto T. Phenylephrine and pilocarpine eye drop test for dementia with Lewy bodies and Alzheimer's disease. *Neurosci. Lett* 2007, 414, 174-177.

Harding GFA, Wright CE, Orwin A. Primary presenile dementia: The use of the visual evoked potential as a diagnostic indicator. *Br. J. Psychol.* 1985, 147, 533-540.

Hinton DR, Sadun AA, Blancks J C, Miller CA. Optic nerve degeneration in Alzheimer's disease. *New Eng. J. of Med.* 1986, 315, 485-488.

Hoenicka J. Genes in Alzheimer's disease. *Revista de Neurolgia* 2006, 42, 302-305.

Hof PR, Morrison JH. Quantitative analysis of a vulnerable subset of pyramidal neurons in Alzheimer's disease. II. Primary and secondary visual cortex. *J. Comp. Neurol.* 1990, 301, 55-64.

Hof PR, Bouras C, Constantinidis J, Morrison JH. Balint's syndrome in Alzheimer's disease: Specific disruption of the occipito-parietal pathway. *Brain Res.* 1989, 493, 368-375.

Imhof A, Kovari E, von Gunten A, Gold G, Rivara CB, Herrmann FR et al. Morphological substrates of cognitive decline in nonagenarians and centenarians: A new paradigm? *J. Neurol. Sci.* 2007, 257, 72-79.

Jellinger KA, Bancher C. Neuropathology of Alzheimer's disease: a critical update. *J. Neural. Transm.* 1998, 54, 77-95.

Johnson AB, Blum NR. Nucleoside phosphatase activities associated with the tangles and plaques of Alzheimer's disease: a histochemical study of natural and experimental neurofibriallary tangles. *J. Neuropath. Exp. Neurol* 1970; 29: 463-478.

Katz B, Rimmer S. Ophthalmologic manifestations of Alzheimer's disease. *Surv. Ophthalmol.* 1989, 34, 31-43.

Kergoat H, Kergoat MJ, Justino L, Chertkow H, Robillard A, Bergman H. Visual retinocortical function in dementia of the Alzheimer type. *Gerontology* 2002, 48, 197-203.

Khatchaturian ZS. Diagnosis of Alzheimer's disease. *Arch Neurol* 1985, 42, 1097-1005.

Kiyosawa M, Bosley TM, Chawluk J et al. Alzheimer's disease with prominent visual symptoms; clinical and metabolic evaluation. *Ophthalmology* 1989, 96, 1077-1085.

Knopman DS. An overview of common non-Alzheimer dementias. *Clinics in Ger Med* 2001, 17, 281.

Lakshminarayanan V, Lagrane J, Kean ML, Dick M, Shankle R. Vision in dementia: contrast effects. *Neurol Res* 1996, 18, 9-15.

Lee G, Cowan N, Kirschner M. The primary structure and heterogeneity of tau protein from mouse brain. *Science* 1988, 239, 285-288.

Lewis DA, Campbell MJ, Terry RD, Morrison JH. Laminar and regional distributions of neurofibrillary tangles and neuritic plaques in Alzheimer's disease. *J. Neurosci.* 1987, 6, 1799-1808.

Mann DMA. The neuropathology of Alzheimer's disease: a review with pathogenetic, aetiological and therapeutic considerations. *Mech. Ageing Dev.* 1985, 31, 213-255.

Mendez MF, Tomsak RL, Remler B. Disorders of the visual system in Alzheimer's disease. *Neurology* 1990, 40, 439-443.

Mirra S, Heyman A, McKeel D et al. The consortium to establish a registry for Alzheimer's disease (CERAD). II. Standardization of the neuropathological assessment of Alzheimer's disease. *Neurology* 1991, 41, 479-486.

Mortimer JA. Alzheimer's disease and senile dementia: Prevalence and Incidence. In: Reisberg B, editor. *Alzheimer's disease: The Standard Reference*. London and New York: MacMillan, 1983, 141-148.

Nakano N, Hatakeyama Y, Fukatsu R et al. Eye-head coordination abnormalities and regional cerebral blood flow in Alzheimer's disease. *Prog in Neuro-Psychopharm and Biol. Psych.* 1999, 23, 1053-1062.

Neargarder SA, Stone ER, Cronin-Golomb A, Oriss S. The impact of acuity on performance of four clinical measures of contrast sensitivity in Alzheimer's disease. *J. Gerontol. Series B* 2003, 58, P54-P62.

Nguyen A, Chubb C, Huff F. Visual identification and spatial location in Alzheimer's disease. *Brain and Cognition* 2003, 52, 155-166.

Nissen MJ, Corkin S, Buoanno FJ, Growden JH, Wray SH, Baver J. Spatial vision in Alzheimer's disease; general findings and a case report. *Arch. Neurol.* 1985, 42, 667-671.

O'Neil D, Rowan M, Abrahams D, Feely JB, Walsh JB, Coakley D. The flash visual evoked potential in Alzheimer type dementia. *Ir. J. Med. Sci.* 1989, 158, 158.

Parisi V, Restuccia R, Fattapposta F, Mina C, Bucci M, Pierelli F. Morphological and functional retinal impairment in Alzheimer's disease patients. *Clin. Neurophysiol.* 2001, 112, 1860-1867.

Parnavelas JG, McDonald JK. The cerebral cortex. In: Emson PC (ed), Chemical Neuroanatomy, Raven Press, New York, 1983, 337-358.

Philpot MP, Amin D, Levy R. Visual evoked potentials in Alzheimer's disease: correlations with age and severity. *Electroenceph. clin. Neurophysiol.* 1990, 77, 323-329.

Prager TC, Schweitzer FC, Peacock LW, Garcia CA. The effect of optical defocus on the pattern electroretinogram in normal subjects and patients with Alzheimer's disease. *Am. J. Ophthalmol.* 1993, 116:, 363-369.

Probst A, Ulrich J, Heitz P. Senile dementia of the Alzheimer type: Astroglial reaction to extracellular tangles in the hippocampus. *Acta Neuropathol* 1982, 57, 75-79.

Rizzo M, Nawrot M. Perception of movement and shape in Alzheimer's disease. *Brain* 1998, 121, 2259-2270.

Roder HM, Eden PA, Ingram VM. Brain protein kinase PK40erk converts tau into a PHF-like form as found in Alzheimer's disease. *Biochem. Biophys. Res. Commun.* 1993, 193, 639-647.

Rogers J, Morrison JH. Quantitative morphology and regional and laminar distributions of senile plaques in Alzheimer's disease. *J. Neurosci.* 1985, 5, 2801-2808.

Rumney NJ. The aging eye and visual appliances. *Opthal. Physiol. Opt.* 1998, 18, 191-196.

Sadun AA, Bassi CJ. Optic nerve damage in Alzheimer's disease. *Ophthalmology* 1990, 97, 9-17.

Sadun AA, Borchert M, DeVita E, Hinton DR, Bassi CJ. Assessment of visual impairment in patients with Alzheimer's disease. *Am. J. Ophthalmol.* 1987, 104, 113-120.

Schlotterer G, Mosovitch M, Crapper-McLachlan D. Visual processing deficits as assessed by spatial frequency contrast sensitivity and backward masking in normal ageing and Alzheimer's disease. *Brain* 1983, 107, 309-325.

Scinto LFM, Daffner KR, Dressler D et al. Potential non-invasive neurobiological test for Alzheimer's disease. *Science* 1994, 266, 1051-1054.

Shipp S, Zeki S. The organization of connections between areas V5 and V2 in macaque monkey visual cortex. *European Journal of Neuroscience,* 1989, 1, 333-354.

Singer W. Central-core control of visual cortex functions. In: Schmitt FO, Worden FG eds. The Neurosciences: 4th Study Program. Cambridge MIT Press, 1979.

Snedecor GW, Cochran WG. Statistical Methods. Iowa State University Press, Ames, Iowa USA, 1980.

Strenn K, Dalbianco P, Weghaupt H, Koch G, Vass C, Gottlob I. Pattern electroretinogram: Luminance electroretinogram in Alzheimer's disease. *J Neural Transm* 1991, Suppl 33, 73-80.

Tales A, Troscianko T, Lush D, Haworth J, Wilcock GK, Butler SR. The papillary light reflex in aging and Alzheimer's disease. *Aging Clin Exp Res* 2001, 13, 473-478.

Tales A, Butler S, Gilchrist I, Jones R and Troscianko T. Visual search in Alzheimer's disease: a deficiency in processing conjunctions of features. *Neuropsychologia* 2002, 40, 1849-1857.

Tierney, M, Fisher, R, Lewis, A et al. The NINCDS-ADRDA work group criteria for the clinical diagnosis of probable Alzheimer's disease. *Neurology* 1988, 38, 359-364.

Trick GL, Barris MC, Bickler-Bluth M. Abnormal pattern electroretinogram in patients with senile dementia of the Alzheimer type. *Ann. Neurol.* 1989, 26, 226-231.

Trick GL, Trick LR, Morris P, Wolf M. Visual field loss in senile dementia of the Alzheimer's type. *Neurology* 1995, 45, 68-74.

Wilhelm B, Wilhelm H, Wormstall H, Kircher T, Kriegbaum C. Dementia of the Alzheimer type and pharmacologic pupil testing. *Nervenheil* 1997, 16, 458-463.

Wood S, Mortel KF, Hiscock M, Bretmeyer BG, Caroselli JS. Adaptive and maladaptive utilization of color cues by patients with mild to moderate Alzheimer's disease. *Arch. Clin. Neuropsychol.* 1997, 12, 483-489.

Zaccara G, Gangemi PF, Muscas GC et al. Smooth-pursuit eye movements: alterations in Alzheimer's disease. *J. Neurol. Sci.* 1992, 112: 81-89.

Zeki SM, Shipp S, The functional logic of cortical connections. *Nature,* 1988, 335, 311-317.

In: Visual Cortex: Anatomy, Functions … ISBN: 978-1-62100-948-1
Editors: J.M. Harris et al. pp. 129-164 © 2012 Nova Science Publishers, Inc.

Chapter 5

A VISUAL COMPUTATIONAL MODEL AND ITS NEURAL COMPUTATION BASED ON INNER PRODUCT IN THE PRIMARY VISUAL CORTEX

***Zhao Songnian*[1], *Zou Qi*[2], *Jin Zhen*[3], *Yao Guozheng*[4], *and Yao Li*[5]*

[1] LAPC, Institute of Atmospheric Physics,
Chinese Academy of Sciences, Beijing 100029, China.
[2] Department of Computer Science,
Beijing Jiaotong University, Beijing 100044, China.
[3] fMRI Center of Brain's function, Beijing 306 Hospital,
Chinese People's Liberation Army, Beijing 100101, China.
[4] College of Information Science, Peking University,
Beijing 100871, China.
[5] State Key Lab. of Recognitive Neuroscience & Learning,
School of Information Science and Technology,
Beijing Normal University,
Beijing 100875, China.

* Corresponding author: yaoli@bnu.edu.cn.

ABSTRACT

Based on the probabilistic reasoning to vision information processing in this paper, combining synchronized response and sparse representation, we propose a new early visual computational model, which consists of the multi-scale filtering, the phase synchronization and the inner product operations. According to features of parallel distributed processing in the visual pathway, the retinal image may be orthogonally divided to sub-unit of image or image patch by means of size of receptive field of ganglion cell, then, all of the information contents contained in image patches transmitted by sub-channel to the primary visual cortex V1, respectively. Further processing to them are made by these distributed cortical functional columns, and its spatial localized, oriented and band-pass characteristics can made response to features of image patches, and from this the realization of inner product operations is achieved, it is an optimal detection operator under the meaning of minimum mean square error to visual image reconstruction. Theoretical analysis and experimental results showed that: at the system level inner product operator reflects the nature of the excitation process of local characteristics of external stimuli onto corresponding neurons in the cortex V1. It is also a plausible assumption about neural computation in V1 cortex. Therefore, it may be have some reference significance to explore the neural mechanisms of the visual information processing. In addition, difficulties, which occur in simulating the receptive field of simple cell in the V1 cortex by sparse coding, are briefly discussed, and the problems, which arise when the cortical functional column is considered as a tiled set of selective filter, also will be compared and analyzed.

Keywords: visual image, primary visual cortex, computational model, phase synchronization, inner product, topological mapping, functional column.

1. INTRODUCTION

Over the past two decades, a lot of research work about the visual information processing carried out in two different directions, and achieved a number of important results, which are, in one direction, the discoveries of synchronous oscillations and synchronized response to visual stimulation in the visual cortex [1-5]; in the other direction, the speculation about the visual cortex to perception of natural scenes, in which sparse representation strategy is used by the visual cortex through exploring the statistical characteristics of natural scenes and its structure [6-13].

In this paper, the relationship between the synchronized response and the sparse representation is researched, to the image processing for vision, the former is essentially a time coding [14-17]; the latter is, on the nature, of space coding [18]. In neuroscience on the vision, both are the embodiment of neurons' receptive field and neural mechanisms of lateral inhibition, are also of an effective representation on retinal images, which embodies following characteristics of visual information processing: high efficiency, robustness and simplicity. In this paper, combining the two, we set up a new computational model of neural network, which consists of multi-scale filtering, phase synchronization and inner product operations and other units.

In this paper, we mainly study the following two processes: one is that the object of the outside world as visual stimulation is transformed into firing spike trains by the ganglion cells, and they then produce synchronized response in the visual cortex, the information about object of the external world contained in the modulated firing spike sequences are demodulated and extracted by the neural phase-locked loops; the other is the information, that is, each primitives of visual image (line segments, corners, contours, etc.) is carry through matches with various receptive fields in different functional columns densely distributed the cortex V1, if a certain orientation and width of image primitives are the same with the orientation and width of the receptive field of some functional column in cortex V1, the corresponding neuron strongly fire [19,20]. Therefore, firing of neurons of functional columns in cortex V1, in this case, is such a firing that is one-to-one corresponding firing, and which reflects activating effect of local features in the visual image to cortical neurons in the cortex V1, is also embodiment of topological connection between retina and cortex V1 as well as visual information and distributed parallel processing [21].

At present, a more popular opinion is that the functional column in the visual cortex is equivalent to a set of tiled filter, which performed filtering processing on the visual image coming from the retina, that is to say, which is to carry out Fourier transform [22], but this transformation is too complicated, too much computation, it seems that it is not suitable for human vision. Instead of point of view about Fourier filtering, we propose a point of view about neural information processing based on topological mapping: one of, cortical functional columns in V1 is no longer seen as a set of filter arrays, but basic image processing unit with the preferential orientation, feature matching. Visual information carried by firing spike trains are transmitted from the retina to the neural phase-locked loop, and are decoded, after that inner product operations of the decoded image signal with functional column of V1 are carried on, the image features contained in primitives are extracted by thousands of cortical function modules in parallel and simultaneously, and then, the visual image is formed in the cortex V1, here only the inner product operation is necessary. Mathematically, the computing complexity of this operation is very low than the Fourier filtering operation [23]. More importantly, the implementation of algorithm is to rely on such a neural mechanism, that is, each visual image unit excited corresponding cortical neurons in order to fire them and is the embodiment of simple function of neurons; and an algorithm's implementation on complex visual image is to rely on a collection operation of a large number of neuron populations based on simple calculation "ON" or "OFF" of each neuron. Furthermore, in theory, that the functional column distributed in the cortex V1 as function module of information processing, its role to visual computing is the inner product, whose realization is based on spatial localized, oriented selection and band-pass characteristics of receptive field of simple cell. When reconstructing visual images, the inner product is also optimal detection operation under the meaning of least mean square error. In order to test the above opinion, we carry on a numerical simulation experiment, in which the receptive field of neurons in the cortex V1 is simulated by using wavelet Gabor pyramid model [24-26], after Lenna picture is pre-filtered by front filter of visual pathway, we get its feature image, and then which is used as the visual test image, the results show that: in this paper, the proposed algorithm with the theoretical prediction is in line, it is some valuable for research the visual information processing from a new light of the perspective.

This paper also briefly discussed difficulties encountered in simulating receptive fields of simple cells in cortex V1 by sparse coding, a variety of natural and man-made images (with their very different statistical characteristics and structure), especially random-dot map and fractals used as the input of computational model simulating receptive field, through multi-scale filtering of front of the visual pathway, these pictures are used to train ICA-based bi-directional neural network, thus we can test the performance and faced difficulties in simulating the outline and shape of the receptive field, which possesses characteristics, such as spatial locality, band-pass property and preferential orientation, we also made a comparative analysis of problem, which arised when the cortical column is considered as a set of tilted selective filter. A brief comparison of our model with the standard model of object recognition (that is, HMAX model [27]) has been made.

2. Computational Model of Vision

The primary visual computational model is as shown in Figure 1, including an array of input light intensity, retina, lateral geniculate nucleus (LGN) and the primary visual cortex (cortex V1), that is, the ventral pathway (so-called "what" pathway [28]). From the perspective of information processing, this is described as below, respectively.

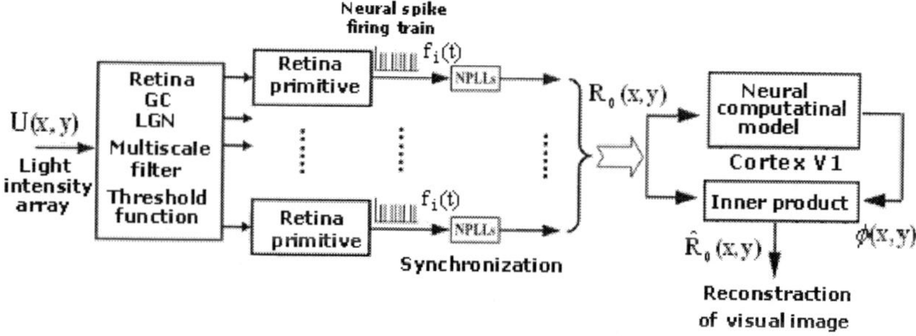

Figure 1. Computational model of the primary vision.

2.1. Multi-scale Representation

In our computational model of visual information processing, we use a neuronal model such as that presented in Figure 2 to simulate visual information preprocessing adopted by photoreceptor cells to bipolar cells to ganglion cells. This neuronal model includes three parts, the linear model for visual input, the threshold transform on the neuron spike firing, and the multi-scale filtering, described below.

Generally, an input image $U_i(x,y)$ is usually expressed as the linear combination of a set of basis function, $U_i(x,y)$, as below

$$U_i(x,y) = \sum_{j=1}^{n} w_{ij} u_j(x,y) - e_i \tag{1}$$

where $u_1, u_2, \cdots, u_n$ are components corresponding to the input lightness intensity array of visual image $U_i(x,y)$, $w_1, w_2, \cdots, w_n$ are weight coefficients among neurons, and e_i is threshold, which is related to whether the neuron fires. In fact, equation (1) is a general form depicting a neuron receiving inputs from several other neurons. In a visual pathway, synaptic connections between photoreceptor cells, bipolar cells, and ganglion cells can be depicted by this equation (1).

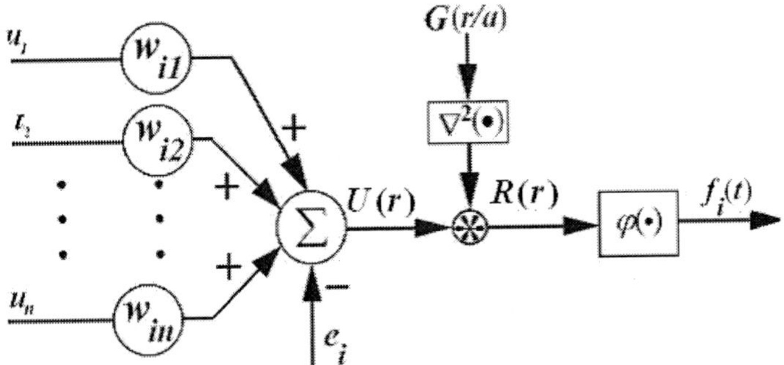

Figure 2. Synaptic connection and neuronal model for visual information preprocessing in the visual pathway. $u_1, u_2, \cdots, u_n$ are inputs from the column matrix of the visual image, $\nabla^2(\cdot)$ is the Laplacian operator, $G(r/\sigma)$ is the two-dimensional Gaussian function, $*$ is a convolution operation, $\varphi(\cdot)$ is the nonlinear threshold function, and e_i is threshold.

On retina ganglion cells receive the input signals coming from photoreceptor cells—bipolar cells and make them into spatial sum, only the ganglion cell whose spatial sum of signals is over the threshold will be able to fire. And many different nonlinear functions may be used as threshold function for different research works. Sigmoid type function (S) is adopted as threshold function in our model.

$$\varphi(v) = \frac{1}{1 + \exp\left(-\sum_{j=1}^{n} [w_{ij} u_j (x, y) - e_i]\right)} \tag{2}$$

Electrophysiological experiments have confirmed that information at multi-scale, such as edges and contours, is transferred by M cells and information at fine scale, such as texture and facial expression, is transferred by P cells [29,30]. Contour information is detected earlier than detailed information so that we can judge an object's category and then recognize the individual within the category. Therefore, multi-scaling properties of receptive fields in ganglion and LGN also should be taken account. In the early 1980s, Marr already gave a clear explanation about neurobiological basis for using the Laplacian operator $\nabla^2 G(r/\sigma)$ to accurately describe the properties. We used this filter to preprocessing for it can simulate ON-center and OFF-center receptive fields of ganglion cells and cells in LGN. Marr discussed plausibility of this filter in information processing in visual pathway [31]. Of course, DoG function (Difference of two Gaussian functions) and Gabor function can also be used as a filter function. It should be mentioned that the filter operator was given by Marr has the following form

$$\nabla^2 G(r/\sigma) = -\frac{1}{2\pi\sigma^4}\left(1-\frac{r^2}{2\sigma^2}\right)e^{-r^2/2\sigma^2} \tag{3}$$

And we are given another operator for the filter

$$\nabla^2 G(r/a) = -\frac{1}{2\pi a^2}\left(1-\frac{r^2}{a^2}\right)e^{-r^2/2a^2} \tag{4}$$

where $r^2 = x^2 + y^2$, σ and a are the scale parameters, which describes the vision can through continuously change the scale to effectively perceive the visual environment and recognize objects for the outside world.

The corresponding spectrum is $\hat{G}(\omega_r)$ and is shown in Figure 3.

$$\hat{G}(\omega_r) = -(\sigma\omega_r)^2 e^{-(\sigma\omega_r)^2/2} \tag{5}$$

The main difference between both filtering operators is: Marr filtering operator does not mean zero, and the mean value of the filtering operator of type (a) is zero (as shown in Figure 4), that is to say, it is an actual wavelet, has also more narrow bandwidth, and can better characterize the actual performance to extract outlines, contours and edges of visual image for visual pre-processing [32].

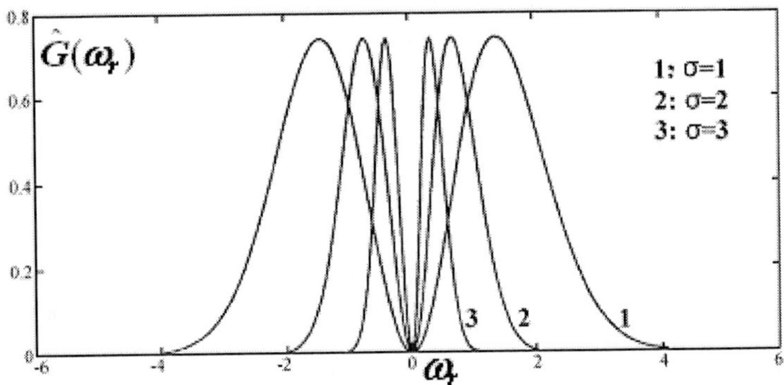

Figure 3. The bandpass filtering property of $\hat{G}(\omega_r)$, ω_r is spatial frequency σ, the scale factor.

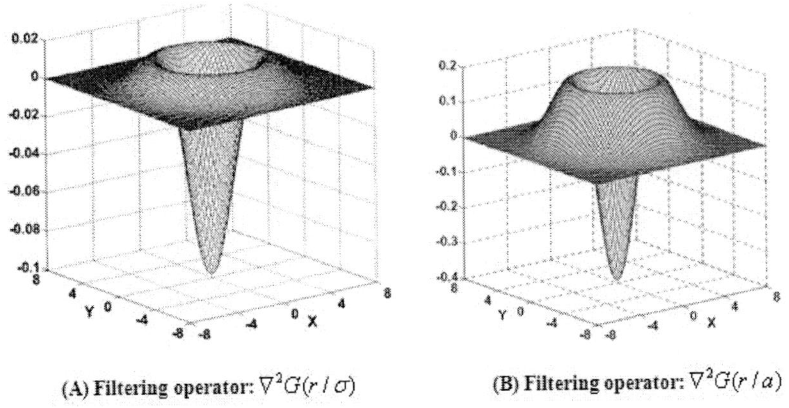

(A) Filtering operator: $\nabla^2 G(r/\sigma)$
(B) Filtering operator: $\nabla^2 G(r/a)$

Figure 4. Comparison of two filtering operators.

In this paper, we use retina images filtered by $\nabla^2 G(r/\sigma)$ as visual images $R(x, y)$

$$R(x, y) = U(x, y) * \nabla^2 G(r/\sigma) \tag{6}$$

$R(x, y)$ can be expressed as the linear superposition of a set of the basic function $\phi_i(x, y)$

$$R(x,y)=\sum_{i=1}^{m}a_i\phi_i(x,y) \tag{7}$$

The operator $\nabla^2 G(r/\sigma)$ can shrink the dynamic range of an input signal's intensity variation, extract the edge, and enhance the main features of the visual image. It not only displays the multi-scaling property of visual information processing but also describes the adaptive functions of visual information processing, such as the constancy and multi-resolution of human vision in perception of the natural scene.

If a visual object is expressed as $f(x)$, where x is a two-dimensional variable, namely $x=(x_1,x_2)$, and its support area D is limited, we have a wavelet transform as follows:

$$\begin{aligned}W_f(a,b)&=\int_{-\infty}^{\infty}f(x)\psi_{a,b}(x)\mathrm{d}x=\int_{-\infty}^{\infty}f(x)\frac{1}{\sqrt{a}}\psi\left(\frac{x-b}{a}\right)\mathrm{d}x\\&=\int_{-\infty}^{\infty}f(ax)\sqrt{a}\psi\left(x-\frac{b}{a}\right)\mathrm{d}x,\ a>0,\ f\in L^2(R),\end{aligned} \tag{8}$$

where $L^2(R)$ denotes a Hilbert space of functions having integral absolute 2-th power; here $f(x)$ is equivalent to $V_0(r)$, and $\psi_{a,b}(x)$ is equivalent to $\nabla^2 G\left(\frac{r-b}{a}\right)$. The multiplication of function $f(x)$ by $\psi_{a,b}(x)$ in formula (7) in nature is a modulated process. In addition, the modulating role of $\psi\left(x-\frac{b}{a}\right)$ to $f(x)$ is equivalent to the modulating role of $\psi\left(x-\frac{b}{a}\right)$ to $f(ax)$; i.e., the role of spreading and shrinkage of $\psi\left(x-\frac{b}{a}\right)$ may be transforming the shrinkage and spreading of $f(ax)$, respectively.

In fact, formula (7) indicates such an essential property of vision—that is, when the scale α increases, $\frac{1}{\sqrt{a}}\psi\left(\frac{x-b}{a}\right)f(x)$ means decreasing the visual acuity ($1/\sqrt{a}$) and increasing the visual contrast $\left[\psi\left(\frac{x-b}{a}\right)\right]$—the vision perceives the external object with support ***D***; $\sqrt{a}\psi(x-b/a)f(ax)$ means increasing visual acuity ($\sqrt{a}$) and decreasing visual contrast $[\psi(x-b/a)]$, and the vision perceives the same external object with the reduced support, D/a. The effects of the abovementioned two cases are the same: both can roughly maintain that the pixel density of an external object is invariable in vision. Therefore, they can reflect the constancy and multi-resolution process of human vision in perceiving the natural scene.

2.2. Synchronized Response in Visual Cortex

Retinal image signal $R(x, y)$, which is topologically mapped along ganglion cells→LGN→V1 , carries and transmits information by encoded neuronal spike trains $f_i(t)$ as shown as in Figure 1, which consists of the phase detecting (PD), the low pass filtering (LP) and the feedback oscillation functions (VCO) .

When we observe external world, neuronal firing spike trains $f_i(t)$ (corresponding to $R(x, y)$) caused by external stimuli (represented by light intensity array corresponding to $U(x, y)$) are random phase signal. Their prior probability is unknown, if we assume it to be uniformly distributed i.e.

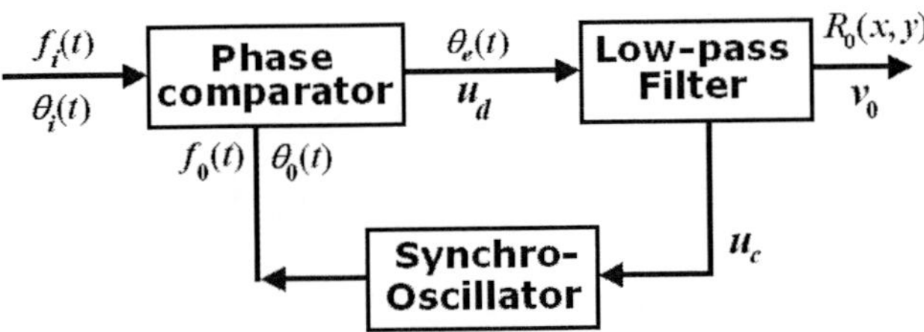

Figure 5. Neural phase-locked loop (NPLL), which consists of phase detector (PD), low-pass loop filter (LP) and voltage controlled oscillator (VCO) . $f_i(t)$ denotes the neuronal firing spike train as the input signal of NPLL; $\theta_i(t)$, $\theta_e(t)$ denotes the input and output phase signals of PC, respectively; $\theta_0(t)$ is the output phase signal of VSO; $u_d(t)$ and $u_c(t)$ denote the input and output voltage signals of LP; and $v_0(t)$ is the output signal of NPLL.

$\theta = 1/2\pi$, according to maximum likelihood approximation we can get

$$\frac{\partial}{\partial \theta} \ln p[f_i(t)/\theta]\Big|_{\theta=\hat{\theta}_{\mathrm{ML}}} = 0 \tag{9}$$

where $p[f_i(t)/\theta] = k\exp\left\{\frac{1}{n_0}\int_0^T [f_i(t) - s(t,\theta)]^2\, dt\right\}$, $\frac{n_0}{2}$ is the power density spectrum of Gaussian white noise. $s(t,\theta)$ is the random phase signal. Generally, we have $s(t,\theta) = A\sin(\omega_0 t + \theta)$, $s(t,\theta)$ is used to simulate synchronous oscillation of neuronal group (at frequency of 30-70Hz). Thus we can obtain

$$\int_0^T f_i(t)\cos(\omega_0 t + \hat{\theta}_{\mathrm{ML}})dt = 0 \tag{10}$$

Many researches have shown that when synchronized response occurs in neural loop [32-35], synchronous oscillation frequency ω_0 will be locked at instantaneous frequency of $f_i(t)$, neuronal spike trains encoded. And all of components of distributed signal $v_i(t)$ are converted into electrical signal. In other words, the signal demodulated from spike trains by neural phase-locked loop will reconstruct a whole visual image $R(x, y)$. At the same time, noises are reduced [35]. Then, subsequent processing is continued in V1.

Incidentally, during the phases of visual image $R(x, y)$ transmitted from ganglion cells to LGN with spike encoding trains and then from LGN to V1, the image signal $R_0(x, y)$ demodulated from spike firing trains by neural phase-locked loop is an accurate copy of $R(x, y)$ without any loss of information. Therefore, tiny difference between $R_0(x, y)$ and $R(x, y)$ can be ignored.

Here, a major role of the NPLL is to demodulate the multi-channel spike firing trains, which contained retinal image information, and to extract primitive image features from these spike firing trains, because only such signals can stimulate the receptive field of cortical neurons in V1, and then make them to activating.

2.3. Optimum Detection in V1

Receptive fields of simple cells in V1 have three features, that is, preferential orientation, band-pass and spatial locality [7,24,25], so they show strong selectivity toward visual image $R(x, y)$. When the image shows properties at specific orientation and specific frequency, corresponding simple cells will make strong response. In other words, when a local patch of image most coincides with receptive field of a simple cell at its sensitive orientation and sensitive frequency, the neuron fires most strongly [19,20]. This can be expressed as

$$r_k = a\phi_k + n_k, \quad k = 1, 2, \cdots, N \tag{11}$$

where r_k is a sample of input, n_k is Gaussian white noise with zero mean and variance σ^2. a is stochastic coefficient. According to maximum likelihood estimation [36]

$$\frac{\partial}{\partial a}\ln p(r/a)\bigg|_{a=\hat{a}_{\mathrm{ML}}} = -\frac{\partial}{\partial a}\sum_{k=1}^{N}\left[\frac{(r_k - a\phi_k)^2}{2\sigma^2}\right]\bigg|_{a=\hat{a}_{\mathrm{ML}}}$$

$$= \frac{1}{\sigma^2}\sum_{k=1}^{N}(r_k - a\phi_k)\phi_k\bigg|_{a=\hat{a}_{\mathrm{ML}}} = 0 \tag{12}$$

By least mean square error rule, the optimum estimation of a is

$$\hat{a}_{\mathrm{ML}} = \frac{\sum_{k=1}^{N} r_k\phi_k}{\sum_{k=1}^{N}\phi_k^2} = \eta\sum_{k=1}^{N} r_k\phi_k \tag{13}$$

where η is a constant. From above derivation, we can see a reaches its optimum value when image sample r_k coincides with receptive fields ϕ_k. It is equivalent to say that receptive fields of simple cells in V1 perform in line with local properties in image $R(x, y)$, which may be called coincidence operation. So, a may be taken as the measure for coincidence operation. To the extremity, a equals 1 when the both of them are completely coincident. In this way, optimal matching is realized between modeling receptive fields ϕ_k and corresponding similar part of image $R(x, y)$. It should be noted that the best matching here is a result in specific resolution or scale, when the resolution or scale changes, matching content, such as contours or details, changes with it, the best match may be the outline, contour of the image also may be some details included in the outline of the image, which is determined by interests of an observer perceives external world. This is just the biological significance of visual multi-scale information processing.

2.4. The Whole Matching between the Retinal Image and Receptive Field Patterns in V1

In section 2.3, we have described the local matching or detection (formula 12). Next we will analyze the image matching as a whole; the purpose of doing so is to obtain a mathematical representation of the neural firing pattern in V1. It is understood that the topological mapping from retinal image $R(x, y)$ to V1 is actually a activating process of $R(x, y)$ with respect to simple selective cells with selectivity to local features of orientation, bandwidth and so on. $G(x, y)$, all of receptive fields of activated neurons population are combined to form the global firing pattern $\Phi(x, y)$ in the cortex, as the responding process of simple cell groups to the visual image. So, the firing process can be regarded as a global matching or coincidence between $G(x, y)$ and $R(x, y)$ at the system level, and eventually responding pattern $\Phi(x, y)$ will be formed. Many methods can be used to measure the extent of matching. However, in order to ensure a minimal reconstruction error, we adopt the following measure:

$$\varepsilon = \iint_B [G(x, y) - R(x, y)]^2 dxdy = \min \tag{14}$$

where B is the imaging region in V1. Thus, we can obtain the following formula

$$\iint_B G(x,y)R(x,y)dxdy \le \sqrt{\iint_B [G(x,y)]^2 dxdy \iint_B [R(x,y)]^2 dxdy} \tag{15}$$

Let $\lambda_{RG} = \iint_B R(x,y)G(x,y)dxdy$, $\lambda_{MAX} = \iint_B R(x,y)^2 dxdy$. In the optimal matching in region B, we will have $R(x,y) = G(x,y)$, which means the mathematical symbol "=" is adopted in formula (9) and $\lambda_{R\Phi}$ reaches its maximum value λ_{MAX}. Therefore, we can define the normalized matching coefficient ρ_{rg} as follows:

$$\rho_{rg} = \frac{\lambda_{RG}}{\lambda_{MAX}} = \frac{\iint_B R(x,y)G(x,y)dxdy}{\iint_B [R(x,y)]^2 dxdy} \tag{16}$$

Obviously when $\rho_{rg}=1$, $G(x,y)$ and $R(x,y)$ reaches a complete match or coincidence. In other words, the receptive fields of all activated neurons in V1 are combined to form the same responding pattern $\Phi(x,y)$ as the whole visual image. It shown that this process can be mathematically described as follows by the inner product

$$\Phi(x,y) = R(x,y) \otimes G(x,y) \tag{17}$$

Additionally, we can see the normalized matching coefficient ρ_{rg} is equivalent to $\hat{a}_{\mathrm{ML}}$ in the previous section.

2.5. Determination of Integral Kernel Function

In the formula (11), the role of whole receptive field $G(x,y)$ of neurons populations in cortex V1 is the same as the integral kernel function in wavelet transform [37]. It can be well described by two-dimensional Gabor wavelet function $G(x,y)_{\lambda,\sigma,\theta,\varphi,\gamma}$ [24,25,38], whose waveform has some local characteristics, such as limited bandwidth and preferential orientation, etc., its expression is as follows

$$G(x,y)_{\lambda,\sigma,\theta,\varphi,\gamma} = \exp\left(-\frac{x'^2+\gamma^2 y'^2}{2\sigma^2}\right)\cos\left(2\pi\frac{x'}{\lambda}+\varphi\right)$$
$$x' = x\cos\Theta + y\sin\Theta$$
$$y' = -x\sin\Theta + y\cos\Theta \tag{18}$$

where γ is the ratio of the length in the major axis direction to that of in minor axis direction, usually set to a constant 0.5; σ is derivative of Gauss, determining the size of receptive fields; φ is the phase, when $\varphi = 0;\pi$; $G(x,y)_{\lambda,\sigma,\theta,\varphi,\gamma}$ is symmetric about the origin; when $\varphi = -(\pi/2);(\pi/2)$ $G(x,y)_{\lambda,\sigma,\theta,\varphi,\gamma}$ is anti-symmetric about the origin; Θ is the optimal orientation, and λ is the wavelength. This waveform should be determined by experimental results from morphology and biophysics, but the exact data are not available so far [39,40]. One plausible way is to set the waveform according to input image features in an input-driven topological mapping [40, 41]. This will be explained in section 3.

Substituting (18) into (17) and considering the cortical responses to orientation and bandwidth properties, we replace $\Phi(x,y)$ with $\Phi(x,y)_{\lambda,\sigma,\theta,\varphi,\gamma}$, we have

$$\Phi(x,y)_{\lambda,\sigma,\theta,\phi,\gamma} = R(x,y)\otimes G(x,y)_{\lambda,\sigma,\theta,\phi,\gamma}$$

$$= R(x,y)\otimes \exp\left(\frac{x'^2+\gamma^2 y'^2}{2\sigma^2}\right)\cos\left(2\pi\frac{x'}{\lambda}+\phi\right) \tag{19}$$

It is the inner product , but not following convolution operation

$$\Phi(x,y)_{\lambda,\sigma,\theta,\varphi,\gamma} = \iint R(u,v)G(x-u,y-v)_{\lambda,\sigma,\theta,\phi,\gamma}\,dudv \tag{20}$$

We know that convolution and cross-correlation operations are essentially filtering operations in the frequency domain, which is not needed for visual information processing in cortex V1, because such a filtering operation would lead to loss of high-and low-frequency information from the retinal picture. The second reason is that the scan process in such operations (convolution and cross-correlation) is a calculation with a high cost (see the section of discussion in this paper, for detail).

From this, we can see that the calculation of the inner product is very well suitable to the visual system in that it satisfies the prerequisites of efficiency, simplicity, and robustness and also provides an optimal means of detection under the condition of least-mean-square-error reconstruction. In fact, formula (20) reflects a specific wavelet transform on retinal image $R(x, y)$ by basis function $G(x, y)_{\lambda,\sigma,\theta,\varphi,\gamma}$ .This formula reflects the neural firing stimulated by the retinal image at the system level. Next we will discuss how to process visual images according to this formula. Two important problems will be discussed, that is, how to divide visual image $R(x, y)$ according to structures and functions of the visual pathway (parallel and multi-channel characteristics) and how to express the orientation selectivity of functional columns in V1 by two-dimensional wavelet function $G(x, y)_{\lambda,\sigma,\theta,\varphi,\gamma}$.

2.6. Division of Visual Image

We know that from the retina to the cortex V1, there are topological mapping by one-to-one, in other words, each basic element (primitives) of visual image is one to one correspondence with the receptive field in corresponding location of cortex V1, the classic neurobiological experiment has shown [19,20], a single neuron is always the most preferentially sensitive to a small part of the specific stimulus, for this reason, the formula (19) can be discretized (see Figure 6) in accordance with the surface size ($a = \Delta x \times \Delta y$) of receptive field of the ganglion cell, and then it is expressed by the relevant matrix.

Typically, the visual image is transmitted to lateral geniculate nucleus by about one million ganglion cells independently and respectively along their own multi-channel, which consist of the optic nerve, and then reached to $4C\alpha$ layer and $4C\beta$ layer of cortex V1 via optic radiation and then the visual image is formed in cortex V1. Clearly, each channel can only send a subset of the image unit, which is an image primitive $r_{i,j}(\Delta x, \Delta y)$. If assuming the number of established channels is $M \times N$,

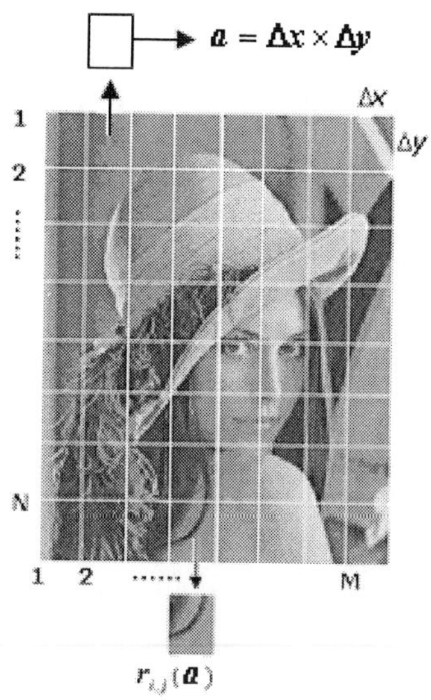

Figure 6. Visual image $R(x, y)$ is divided into $M \times N$ local patches according to a ganglion cell's receptive field.

which means that visual image should has been divided into $M \times N$ sub-units (see Figure 6), the size of each sub-unit of the image can be also assumed that are the same with size of receptive field of ganglion cells, so that, the image transmitted by each channel is only a subset of units (that is, the image element), which contains local features of an image, the image array $[R_{i,j}(\Delta x, \Delta y)]_{M \times N}$ formed by the orderly spatial sum of all the sub-unit can be expressed as the following matrix:

$$[R_{i,j}(\Delta x, \Delta y)]_{M \times N}$$

$$= \begin{bmatrix} r_{1,1}(\Delta x, \Delta y) & r_{1,2}(\Delta x, \Delta y) & \cdots & r_{1,N}(\Delta x, \Delta y) \\ r_{2,1}(\Delta x, \Delta y) & r_{2,2}(\Delta x, \Delta y) & \cdots & r_{2,N}(\Delta x, \Delta y) \\ \vdots & \vdots & \vdots & \vdots \\ r_{M,1}(\Delta x, \Delta y) & r_{M,2}(\Delta x, \Delta y) & \cdots & r_{M,N}(\Delta x, \Delta y) \end{bmatrix}$$

$$i = 1,2,\cdots,M,\ j = 1,2,\cdots,N. \tag{21}$$

Similarly, if the receptive field of a neuron distributed in V1 is denoted $g_{i,j}(\Delta x, \Delta y)$, these neurons activated by image array $[R_{i,j}(\Delta x, \Delta y)]_{M\times N}$ and then will form a response pattern $[\Phi_{i,j}(\Delta x, \Delta y)]_{M\times N}$ as the pattern of cortical receptive field, can be also expressed as a $M \times N$ matrix as follows:

$$[\Phi_{i,j}(\Delta x, \Delta y)]_{M\times N} =$$
$$\begin{bmatrix} \phi_{1,1}(\Delta x, \Delta y) & \phi_{1,2}(\Delta x, \Delta y) & \cdots & \phi_{1,N}(\Delta x, \Delta y) \\ \phi_{2,1}(\Delta x, \Delta y) & \phi_{2,2}(\Delta x, \Delta y) & \cdots & \phi_{2,N}(\Delta x, \Delta y) \\ \vdots & \vdots & \cdots & \vdots \\ \phi_{M,1}(\Delta x, \Delta y) & \phi_{M,2}(\Delta x, \Delta y) & \cdots & \phi_{M,N}(\Delta x, \Delta y) \end{bmatrix}$$
$$i = 1,2,\cdots,M,\ j = 1,2,\cdots,N. \tag{22}$$

According to neurophysiology and neuroanatomy [28,41-45], cortical modules are densely distributed in V1, with approximately 10^3 modules; the area of each module is approximately 1.8 mm×1.8 mm, containing two functional columns for both left and right eyes. Thus, the area related with every functional column $B_{k,l}(s)$ is 0.9 mm×0.9 mm. At the system level, at present, we also do not have enough knowledge about neurophysiology and neuroanatomy of columns, it is reasonable to assume that these functional columns have the same information processing function and are composed of many receptive fields with different orientations and frequencies [28,41].

In this paper, the receptive fields of the functional column $g_{i,j}(\Delta x, \Delta y)$ can be represented as a matrix. As in Figure 6, each row of the matrix represents eighteen oriented receptive fields of the same type uniformly distributed from 0° to 180°. Each column of the matrix represents eight types of receptive fields with a same orientation, which consist of Gabor orthogonal function with different frequencies. (When more detailed description, of course, it also may be available to increase the number of receptive field). So, $[B_{k,l}(s)]_{K\times L}$ is made up of 144 elements $g_{i,j}(b)$ ($k = 1,2,\cdots,K$; $l = 1,2,\cdots,L$) as shown as Figure 7.

$$B_{k,l}(s) = [B_{k,l}(s)]_{K\times L} =$$

$$= \begin{bmatrix} g_{1,1}(\Delta x, \Delta y) & g_{1,2}(\Delta x, \Delta y) & \cdots & g_{1,18}(\Delta x, \Delta y) \\ g_{2,1}(\Delta x, \Delta y) & g_{2,2}(\Delta x, \Delta y) & \cdots & g_{2,18}(\Delta x, \Delta y) \\ \vdots & \vdots & \cdots & \vdots \\ g_{8,1}(\Delta x, \Delta y) & g_{8,2}(\Delta x, \Delta y) & \cdots & g_{8,18}(\Delta x, \Delta y) \end{bmatrix} \quad (23)$$

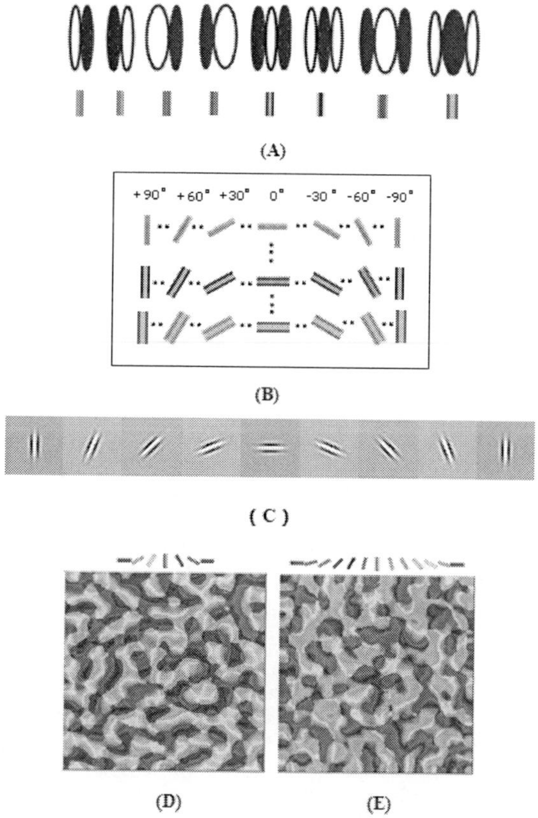

(A)

(B)

(C)

(D) (E)

Figure 7. Functional columns as basic information processing units

(a) eight representative types of receptive fields in functional columns in V1; (b) orientations range from 0° to 180° with a same interval of 10°; (c) a type of receptive field calculated by formula (12); (d) actually measured in experiment and (e) simulated pinwheel-like structure of cortical functional column of rhesus monkey [41].

Therefore, some edge or a part of contour with area a located at (i, j) in retinal image $[R_{i,j}(\Delta x, \Delta y)]_{M\times N}$ can find best match with the receptive field $g_{i,j}(\Delta x, \Delta y)$ with the same orientation and shape. When $\rho_{r\phi} = 1$, it means the patch at (i, j) in $[R_{i,j}(\Delta x, \Delta y)]_{M\times N}$ completely matches the cortical module $[B_{k,l}(s)]_{K\times L}$ with special orientation. Then the neuron is activated and it will produce strongest responses, the diagram of this principle of image reconstruction method shown in Figure 8.

When all the $M \times N$ patches in retinal image $[R_{i,j}(\Delta x, \Delta y)]_{M\times N}$ simultaneously parallel activate corresponding neurons in topological way, they form response pattern $[\Phi_{i,j}(\Delta x, \Delta y)]_{M\times N}$ in V1.At the system level, this process can be described by inner product in Hilbert space.

Optimal responding of a neuron in V1 is actually feature detection of cortical module $[B_{k,l}(s)]_{K\times L}$ to each image patch $r_{i,j}(\Delta x, \Delta y)$ in retinal image $[R_{i,j}(\Delta x, \Delta y)]_{M\times N}$, which can be represented as product of them, i.e. $r_{i,j}(\Delta x, \Delta y)[B_{k,l}(s)]_{K\times L}$, its discretized form can be expressed as following matrix form:

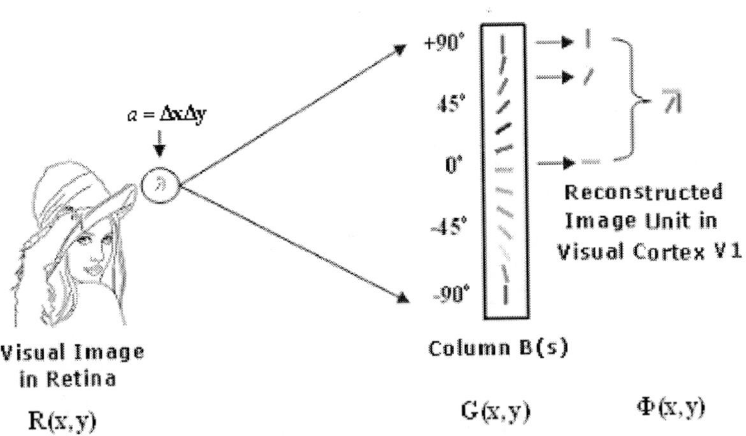

Figure 8. Optimal matching between an image patch (upper right corner of the Mexico hat) and receptive fields of specific orientations in cortical modules $[B_{k,l}(s)]_{K\times L}$

$$[\Phi_{i,j}(\Delta x,\Delta y)]_{M\times N} \square [R_{i,j}(\Delta x,\Delta y)]_{M\times N} \otimes B_{k,l}(s) =$$

$$= \begin{bmatrix} r_{1,1}(\Delta x,\Delta y) & r_{1,2}(\Delta x,\Delta y) & \cdots & r_{1,N}(\Delta x,\Delta y) \\ r_{2,1}(\Delta x,\Delta y) & r_{2,2}(\Delta x,\Delta y) & \cdots & r_{2,N}(\Delta x,\Delta y) \\ \vdots & \vdots & \vdots & \vdots \\ r_{M,1}(\Delta x,\Delta y) & r_{M,2}(\Delta x,\Delta y) & \cdots & r_{M,N}(\Delta x,\Delta y) \end{bmatrix} \otimes \left[B_{k,l}(s) \right] =$$

$$= \left. \begin{bmatrix} r_{1,1}(\Delta x,\Delta y)B_{k,l}(s) & r_{1,2}(\Delta x,\Delta y)B_{k,l}(s) & \cdots & r_{1,N}(\Delta x,\Delta y)B_{k,l}(s) \\ r_{2,1}(\Delta x,\Delta y)B_{k,l}(s) & r_{2,2}(\Delta x,\Delta y)B_{k,l}(s) & \cdots & r_{2,N}(\Delta x,\Delta y)B_{k,l}(s) \\ \vdots & \vdots & \vdots & \vdots \\ r_{M,1}(\Delta x,\Delta y)B_{k,l}(s) & r_{M,2}(\Delta x,\Delta y)B_{k,l}(s) & \cdots & r_{M,N}(\Delta x,\Delta y)B_{k,l}(s) \end{bmatrix} \right|_{\max}$$

$$\text{for} \begin{cases} \forall \left\{ r_{i,j}(\Delta x,\Delta y)B_{k,l}(s) \right\} \\ \hat{a}_{\mathrm{ML}} = 1, \text{or} \rho_{r\varphi} = 1 \end{cases} \qquad (24)$$

where $\otimes$ denotes inner product [33,41]. The symbol $|\max$ for $\forall \left\{ r_{i,j}(\Delta x,\Delta y)B_{k,l}(s) \right\}$, $\hat{a}_{\mathrm{ML}} = 1$, or $\rho_{r\phi} = 1$ denotes the maximal of all products between $r_{i,j}(\Delta x,\Delta y)$ and $B_{1,1}(s), B_{1,2}(s), \cdots,$ $B_{8,18}(s)$ .The neurobiological meaning is that only stimuli with optimal orientation and frequency can activate strongest response of simple cells in V1. Formula (24) represents the activating pattern corresponding to a typical image. It can be represented as matrix

$$[\Phi_{i,j}(\Delta x,\Delta y)]_{M\times N} =$$

$$= \begin{bmatrix} \left\{ r_{1,1}(\Delta x,\Delta y)B_{k,l}(s) \right\} \big|_{\max} & \left\{ r_{1,2}(\Delta x,\Delta y)B_{k,l}(s) \right\} \big|_{\max} & \cdots & \left\{ r_{1,N}(\Delta x,\Delta y)B_{1,N}(s) \right\} \big|_{\max} \\ \left\{ r_{2,1}(\Delta x,\Delta y)B_{k,l}(s) \right\} \big|_{\max} & \left\{ r_{2,2}(\Delta x,\Delta y)B_{k,l}(s) \right\} \big|_{\max} & \cdots & \left\{ r_{2,N}(\Delta x,\Delta y)B_{k,l}(s) \right\} \big|_{\max} \\ \vdots & \vdots & \cdots & \vdots \\ \left\{ r_{M,1}(\Delta x,\Delta y)B_{k,l}(s) \right\} \big|_{\max} & \left\{ r_{M,2}(\Delta x,\Delta y)B_{k,l}(s) \right\} \big|_{\max} & \cdots & \left\{ r_{M,N}(\Delta x,\Delta y)B\ (s) \right\} \big|_{\max} \end{bmatrix}$$

(25)

In formula (19), $[\Phi_{i,j}(\Delta x,\Delta y)]_{M\times N}$ is representation of retinal image $R(x,y)$ in V1. It has essential difference with traditional coding.

3. Numerical Simulation

In numerical simulations, according to resolution of $10^\circ/50\mu m$, we divide orientations of receptive fields in functional columns into 18 parts ranged from 0° to 180° by every 10° interval, i.e. 10°, 20°, $\cdots$, 180°. Eight types of receptive fields are calculated by Gabor function $G(x,y)_{\lambda,\sigma,\theta,\varphi,\gamma}$ according to formula (24) and form arrays $B_{k,l}(s)$ according to formula (23). Then the test image Lenna is divided into $M \times N$ patches according to formula (25). Activating pattern of every receptive field $\phi_{i,j}(\Delta x, \Delta y)$ stimulated by a patch $r_{i,j}(\Delta x, \Delta y)$ is calculated by formula (21). The whole activating pattern $[\Phi_{i,j}(\Delta x, \Delta y)]_{M\times N}$ stimulated by $[R_{i,j}(\Delta x, \Delta y)]_{M\times N}$ is processed by formula (21) and (21, 22).

Considering multi-scale property of visual system, we introduce a factor $2^k a$ to reflect scaling role of cortex to retinal images. Then $[\Phi_{i,j}(\Delta x, \Delta y)]_{M\times N}$ in formula (21) can be replaced by $[\Phi_{2^k a}(m,n)]_{M\times N}$ where $a = \Delta x \times \Delta y$, k=0, 1,2. It can be represented by matrix

$$[R_{2^k a}(m,n)]_{M\times N} =$$

$$= \begin{bmatrix} \{r_{2^k a}(1,1)B_{k,l}\}|_{\max} & \{r_{2^k a}(1,2)B_{k,l}\}|_{\max} & \cdots & \{r_{2^k a}(1,N)B_{k,l}\}|_{\max} \\ \{r_{2^k a}(2,1)B_{k,l}\}|_{\max} & \{r_{2^k a}(2,2)B_{k,l}\}|_{\max} & \cdots & \{r_{2^k a}(2,N)B_{k,l}\}|_{\max} \\ \vdots & \vdots & \cdots & \vdots \\ \{r_{2^k a}(M,1)B_{k,l}\}|_{\max} & \{r_{2^k a}(M,2)B_{k,l}\}|_{\max} & \cdots & \{r_{2^k a}(M,N)B_{k,l}\}|_{\max} \end{bmatrix} \tag{26}$$

According to formula (23), the forms of V1 cortical neurons' receptive field of the 3 types as shown in Figure 9. 18 of their respective field have different direction, different bandwidth and different localized characteristics, respectively.

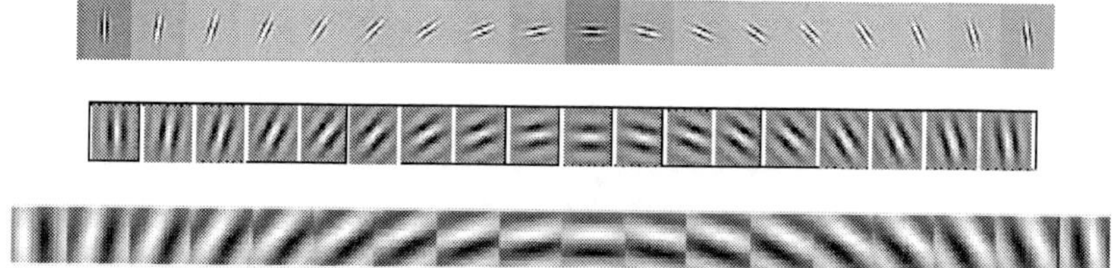

Figure 9. Three types of receptive fields in function columns in V1 calculated by Gabor. Orientations are 0°, 10°, 20° , ⋯, and 180° in turn.

The result of numerical simulation is given in Figure 10.

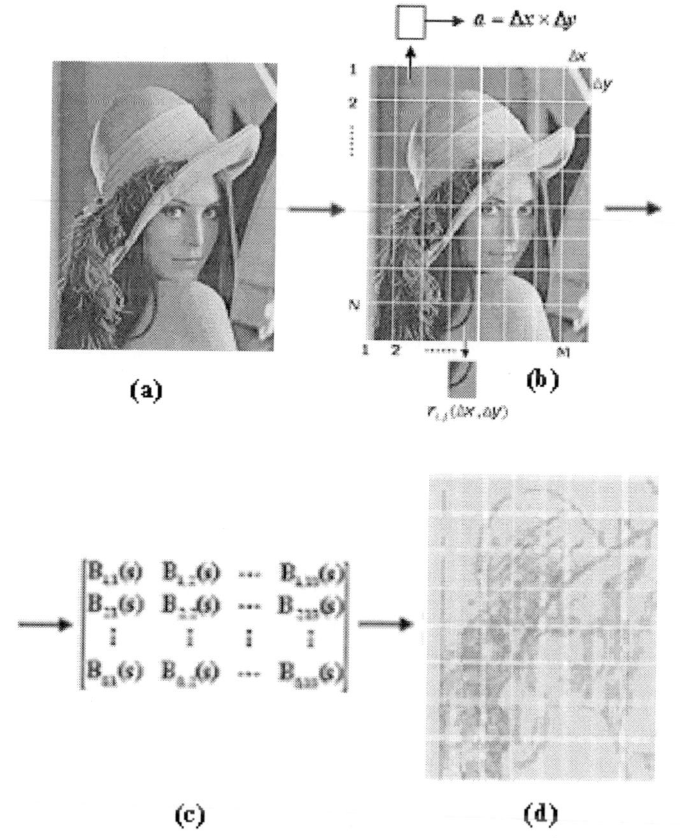

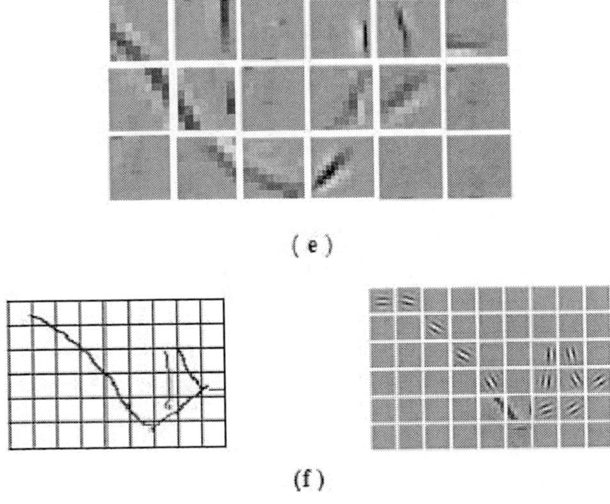

Figure 10. Image reconstruction by topological mapping and inner product (a) source image Lenna, (b) retinal image $[R_{i,j}(\Delta x, \Delta y)]_{M\times N}$, (c) $B_{k,l}(s)$ array of functional columns in cortex V1, (d) reconstruction of Lenna image, (e) enlargement of visual image, (f)a part of activating pattern $[\Phi_{i,j}(\Delta x, \Delta y)]_{M\times N}$ in cortex V1 calculated by formula (18) (upper right corner of the Mexico hat).

4. DISCUSSION

Visual information processing model includes information processing of the primary visual cortex V1, therefore, it is a very important issue to understand how to deal with the visual image come from retina by distributed receptive fields of simple cells in the cortex V1, and its property localized, the orientation selective and band-pass characteristic. In other words, it is what kind of neural computation, how to express mathematically, is a very important issue.

On the present situation, our understanding of these issues is also very small. However, the current research about neurophysiologic and anatomic functions of cortical columns in the cortex V1, especially the exploration on its information processing characteristics, there have been some preliminary results [39], which involves information processing mechanisms of V1 cortical column and implementation of algorithms, that is, the meaning of formulae (19) and (24) in this paper.

Currently, the popular opinion is that simple cells densely distributed in V1 have similar function with a tiled set of selective spatio-temporal filters, while function of cortex V1 is similar to local, complex Fourier transform. Theoretically, it can realize many neural processing about frequency, orientation, motion and other spatio-temporal operations [40-45].

In this opinion, responding property of $\Phi(x,y)_{\lambda,\sigma,\theta,\varphi,\gamma}$ in cortex V1 is achieved by convolution $(*)$ between image $R(x,y)$ and receptive fields $G(x,y)_{\lambda,\sigma,\theta,\varphi,\gamma}$ [22].

$$\begin{aligned}\Phi(x,y)_{\lambda,\sigma,\theta,\varphi,\gamma} &= R(x,y) * G(x,y)_{\lambda,\sigma,\theta,\varphi,\gamma} \\ &= \iint R(x,y)G(x-u,y-v)_{\lambda,\sigma,\theta,\varphi,\gamma}\,dudv \end{aligned} \tag{27}$$

That's to say, we should take $G(x,y)_{\lambda,\sigma,\theta,\varphi,\gamma}$ as template and scan the whole image $R(x,y)$ from top to bottom and from left to right. For example, if $G(x,y)_{\lambda,\sigma,\theta,\varphi,\gamma}$ is a rod-like receptive field with a tilt of 45° to the right, it will match to many edges with similar orientation in $R(x,y)$, so many cells in cortex V1 are activated. The activating pattern $\Phi(x,y)_{\lambda,\sigma,\theta,\varphi,\gamma}$ is given in Figure 11 (a). Similar activating pattern corresponding to a vertical edge is given in Figure 11 (b). Therefore, this is not an effective method, because it stimulates too many responses of relative cells and costs energy greatly [23]. On the contrary, the product operation in the inner product algorithm is equivalent to the activation process that means the image primitives evoke neuron with corresponding receptive field. Obviously, this calculation cost is very small, and therefore, it is more in line with the multi-channel parallel processing mechanism of biological vision.

In order to compare computational complexity of two methods, we express formula (27) in the following discrete form

$$\begin{aligned}\phi(i,j) &= \sum_{m=1}^{M}\sum_{n=1}^{N} r(m,n)g(i-m,j-n) \\ i &= 1,2,\cdots,M;\ j=1,2,\cdots,N\end{aligned} \tag{28}$$

every element $\phi(i, j)$ in array $[\Phi_{i,j}(\Delta x, \Delta y)]_{M\times N}$ must be counted $M \times N$ times, the total counted times for all elements is $M^2 \times N^2$.

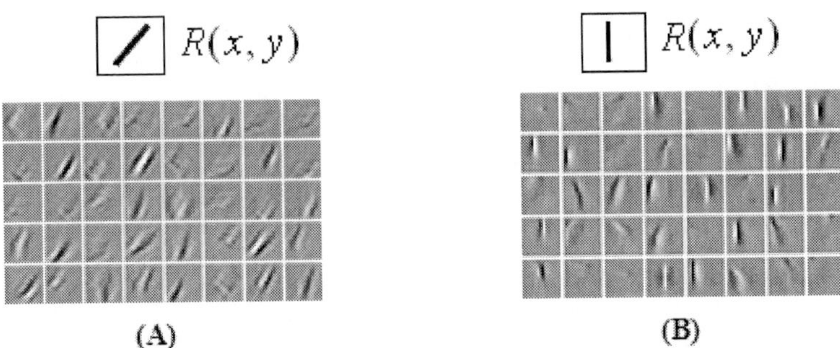

Figure 11. Results of convolution operation stimuli with receptive fields in visual image $R(x,y)$ (a) stimuli is tilted edge, (b) stimuli is vertical edge.

While in our method, in order to reconstruct retinal image $[R_{i,j}(\Delta x, \Delta y)]_{M\times N}$, we only calculate activating pattern $\phi_{i,j}(\Delta x, \Delta y) = r_{i,j}(\Delta x, \Delta y)B_{k,l}(s)$ of receptive field stimulated by every patch $r_{i,j}(\Delta x, \Delta y)$ according to formula (13), and then determine location of every patch topologically mapped to V1 according to formula (24) and (25). Finally we get the whole activating pattern $[\Phi_{i,j}(\Delta x, \Delta y)]_{M\times N}$ stimulated by image $[R_{i,j}(\Delta x, \Delta y)]_{M\times N}$. Obviously, computational complexity of our method is low, so it more accords with multi-channel parallel processing mechanism in biology vision.

In our method, main computation is $\phi_{i,j}(\Delta x, \Delta y) = r_{i,j}(\Delta x, \Delta y)B_{k,l}(s)$, so the counted times are $M \times N \times K \times L$ ($K << M, L << N$). We can see computational complexity of inner product is much lower than convolution.

For sparse coding, which focus on the V1 cortex how to express the visual image at least the number of cells, it is based on the constraint of energy function minimization to construct a class of neural network learning model without supervisor, which can simulate shape and distribution of receptive field of simple cell in cortex V1, its results are "similar" to which determined by electrophysiological experiments, and because of this, further speculation is made, that is, the visual cortex V1 adopts sparse coding strategy to process visual information.

The neural network models established using different learning algorithm mainly are Simoncelli's wavelet transform algorithm [46]; Olshausen and Field' over complete basis algorithm [47]; Hyvarinen, Oja and Hoyer' independent component analysis (ICA) algorithm [48,49] and so on. In general, ICA algorithms is higher efficient, and it's bilateral network has high accuracy in performance of image reconstruction, therefore, when simulating receptive field of cortical simple cell, ICA algorithm is often used. However, all of these algorithms are too complicated algorithm, and the order of receptive field of cortical simple cell given by these algorithms is random and unordered for each time, obviously it does not conform to the orderly arranged orientation of visual cortex functional columns.

We used a high-resolution picture of Lenna, a picture filtered using a multi-scale filter, and a line drawing (see Figure 12 (a), (b), and (c)) to train an ICA network, respectively. The modeled receptive field obtained from the ICA network is shown in Figure 12(d), (e), and (f), respectively. There are obvious differences between the three modeled perceptive fields, it can be see that they have different display order and shapes, even though the three images used to train the ICA network have the same contour and edge features. The results indicate that the algorithms above are not those adopted by biological vision.

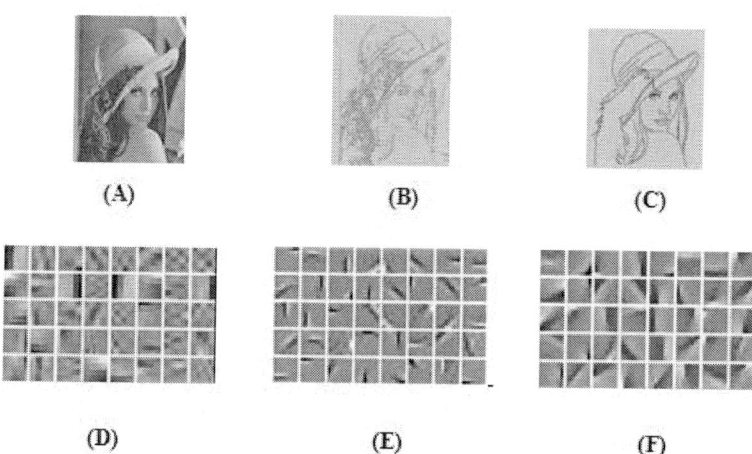

Figure 12. Three different receptive fields obtained by ICA.

(a) A high-resolution image of Lenna; (b) image of Lenna filtered by the operator $\nabla^2 G(r/\sigma)$; (c) line drawing of Lenna; (d), (e) and (f) are receptive fields corresponding to (a), (b) and (c), respectively.

If the random-dot diagram will be used to training computational network of sparse coding, it is difficult to form a pattern of receptive field, as shown in Figure 13.

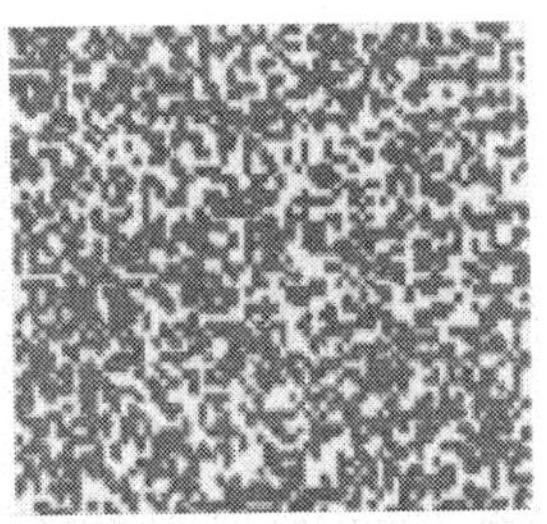

(a) random dot image

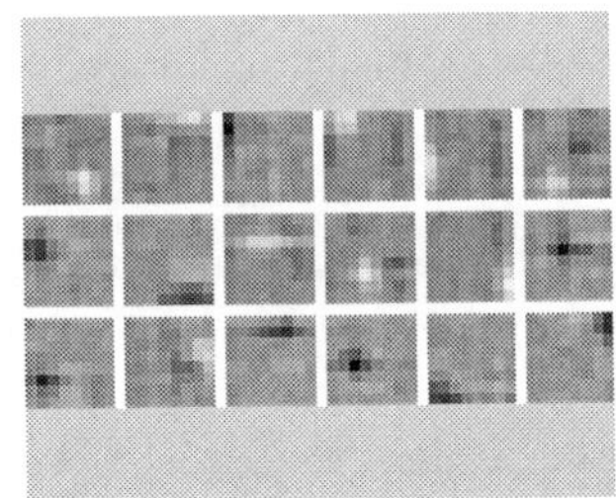

(b) receptive fields of random dot image

Figure 13. modeling receptive fields using random dot image as training picture.

Recently, after the algorithm of sparse coding has been improved [50], it can simulate a greater number (34 × 34) of receptive fields in cortex V1, ordering and sorting of receptive fields has been also improved. If the research in this area can be used to simulate a more realistic distribution mode of receptive field, when establishing the neural network model of visual information processing, it no longer to rely on formula (18) to calculate the shape of receptive field, and then to carry out convolution by formula (28) or inner product operation by formula (25), but it is available to directly deal with visual image using simulated receptive field of neural network, this will have important meaning to understanding of the information processing mechanism of cortical functional columns.

5. A Brief Summary and Conclusions

The rapid development and acquired important results in visual neuroscience has been to lay the foundation for establishing neural computational network to explore visual information processing, and the computational theory about vision has already provided the basis for testing neural network models. At present, significant developments on the visual information processing has been made in two aspects, one of which is synchronized response, which has contributed to the problem of integration of distributed information processing; the other is sparse coding of simple cells in visual cortex, which has effectively simulated towards distributed features of receptive fields of simple cell in visual cortex, such as space localized, band-pass characteristic and orientation selective, and in this paper, we combine the two to propose a new neural network computational model of visual information processing, which consists of the multi-scale filtering, phase synchronization, and inner product operation. Theoretical analysis and numerical simulated results show that this model can better reflect information processing characteristics of human visual system, namely: simplicity, efficiency and robustness. Therefore, they are to have an important reference value to the current study of the visual neural computation.

We know that the visual image in the retina must be projected onto the visual cortex V1 by one-to-one, the follow-up of all higher cortical processing have access to information from the cortex V1. It is clear that in the cortex V1, the loss of any information is not allowed, each element of the visual image in the retina should be correspond to each element of the receptive field in the cortex V1 by one-to-one, and only in this way can the cerebral cortex precisely perceive the visual image of the retina. Neurophysiology have indicated that retinal images are topologically projected onto the visual cortex, each corresponding neurons in more than threshold will be excited and are in activating state. The overall activity pattern of neurons' receptive field in cortex V1 will be high-fidelity reproduction of the retinal image. In other words, the total features in a image patch, such as the line segment with some direction, or a part of contour lines in image primitives can be projected onto its corresponding cortical neuron, then, the optimum matching between them is achieved, the neuron is activated.

From the viewpoint of signal processing, that the multiplication, $r_{i,j}(\Delta x, \Delta y)B_{k,l}(s)$, shows that receptive field $g_{i,j}(\Delta x, \Delta y)$ of cell in cortex V1 is excited by the corresponding retinal image $r_{i,j}(\Delta x, \Delta y)$, and the cell will be in activated state. Thus, such operations in line with the mechanism of neurobiology, it is both the embodiment of simple functions of neurons and reflection of collective operation of a large number of neuron population, such collective operation is only peculiar to advantages of biological systems. The $\phi_{i,j}(\Delta x, \Delta y)$ is local activated pattern of the visual cortex V1, which corresponding to image primitive $r_{i,j}(\Delta x, \Delta y)$ .Different visual stimulation images will produce different activated mode of cortical neuron groups, the activated state caused by only the detail of visual stimulus images is much weaker than that caused by the outline, edge and contour in stimulus image, this has been confirmed by results about sparse coding theory [51].

Visual information processing in V1 is very important, but our understanding at the system level remains litter since recent 30 years from Hubel and Wiesel's discover [19,20] in 1960s to Field and Olshausen's sparse coding theory [10] in 1990s. Especially, our understanding to the mapping relations and the pinwheel-like distributed structure of preferential orientation, spatial frequency characteristics of functional columns in visual cortex, and its role in visual image processing is very litter [52,53]. Therefore, whether our neural computation model based on current existing knowledge about structure and function of functional column in visual V1 [54-60] is a reasonable and natural description to need further test in neurobiology.

APPENDIX

In this article proposed visual computational model is shown in Figure 14.

The main purpose that we proposed the model is to study what kind of neural computation is used by visual cortex (V1) to process the retinal image, and mathematically we make an attempt to provide a proper description according to the model, which is different with those models whose main purpose are lie in how to perceive or recognize object, there are many such kind of models, representative and well-known models are HMAX model of object recognition proposed by Riesenhuber and Poggio in 1999 [27], attention-driven model of perception and recognition proposed by Rybak et al in 1998 [61], and so on. The model in this paper just only related to the cortex V1, but these hierarchal cognitive models are related to information processing of middle and high cortical areas of human. Only in terms of V1 cortex, the essential differences between these two types of models are in: the visual cortex V1 how to represent (or processing) visual image. The former is based on formula (18), namely inner product operations; the latter is based on formula (20) that is convolution operation. In Figure 9, it has been pointed out that convolution operation will activate overmuch cortical neurons, and so that they are in active state (i.e., one single input yield multiple output), in order to overcome this shortcoming in HMAX model, Riesenhuber and Poggio used nonlinear maximum operation to identify maximum output of simple cells in cortex V1 (corresponding to simple cells' postsynaptic potential) from the convolution operation, then, make it as the input of complex cells, due to the complex cell is not sensitive to the position of the stimulus image in its receptive field, thus, the output of complex cell is invariant to position.

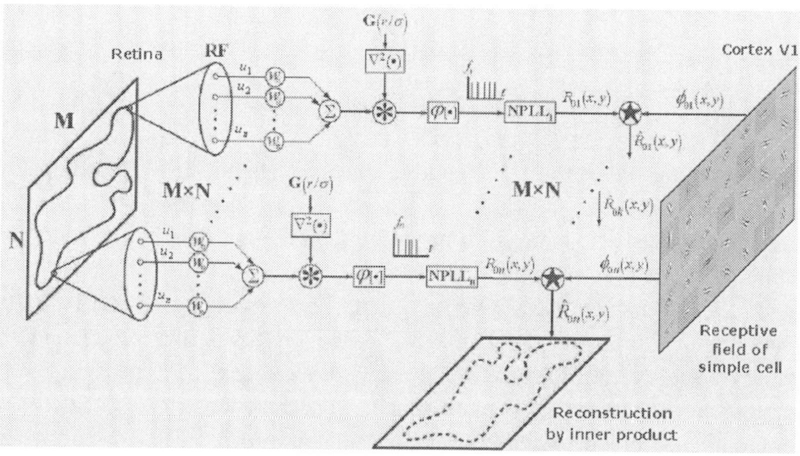

Figure 14. Computational model of visual information processing in the visual pathway.

RF, receptive field; $u_1, u_2, \cdots, u_n$, inputs of the visual image; $G(r/a)$, two-dimensional Gaussian function; $\nabla^2(\cdot)$, Laplacian operator; $*$, convolution operation; $\varphi[\cdot]$, nonlinear threshold function; NPLL, neural phase-locked loop. $R_0(x,y)$ is a precise replica of the external simulation $u(x,y)$; $\hat{R}_0(x,y) = R_0(x,y) \bigstar \phi(x,y)$ is the inner product between $R_0(x,y)$ and $\phi(x,y)$, the receptive field of a simple cell in the visual cortex V1; $\bigstar$, is the inner-product operation. In this example, the Lenna image at high resolution is the input signal $u(x,y)$, and the output $\hat{R}_0(x,y)$ of the whole visual-computation model is a copy of the Lenna image formed by activated simple cells in V1.

ACKNOWLEDGEMENTS

This work is supported by NSFC of China (60371045, 60628101,60805040,and 90820019). Ma Xiaoguang had drawn all these

Figures. The first author deeply appreciates his kind help, and we would like to thank the anonymous reviewers for their detailed comments and questions which improved the quality of our manuscript.

NOTE

Competing interests: The authors have declared that no competing interests exists.

Part of this article is taken from the literatures published by the following journals. The authors are grateful to them.

1. Zhao Songnian, Zou Qi, Jin Zhen, Yao Guozheng, Yao Li (2010), Neural computation of visual imaging based on Kronecker product in the primary visual cortex , *BMC Neuroscience*, 11:43: 1-13.
2. Zhao Songnian, Zou Qi, Jin Zhen, Yao Guozheng, Yao Li (2010), A Computational model of early vision based on synchronized response and inner product operation, *Neurocomputing* 73, 3229–3241.
3. Zhao Songnian, Xiong Xiaoyun, Yao Guozheng, Fu Zhi (2003), A Computational model as neurodecoder based on synchronous oscillation in the visual cortex, *Neural Computation*, 15,2399-2418

REFERENCES

[1] Deadwyler SA, Hampson RE. Ensemble activity and behavior: what's the code? *Science*, 1995, 270, 1316-1318.

[2] von der Malsburg C. Am I Thinking Assemblies? In: Pem, G. & Aersten, A, eds, Brain Theory, Berlin, Springer-Verlag, 1986, 161-176.

[3] Gray CM, Singer W. Stimulus-specific neuronal oscillations in orientation columns of cat visual cortex. *Proc Natl Acad Sci USA*, 1989, 86, 1698-1702.

[4] Gray CM, Konig P, Engel AK, et al. Oscillatory responses in cat visual cortex inter-columnar synchronization which reflect global stimulus properties. *Nature*, 1989,338, 334-337.

[5] Eckhom R, Basar R, Jordan W, et al. Coherent oscillation : a mechanism of feature linking in the visual cortex ? Multiple electrode and correlation analyses in the cat. *Biol Cybern*, 1988, 60, 121-130

[6] Field DJ. Relations between the statistics of natural images and the response properties of cortical cells, *Opt Soc Am*, 1987, A, 4, 2379-2394.

[7] Field DJ. Scale-invariance and self-similar 'wavelet' transforms : an analysis of natural scenes and mammalian visual systems. In : Wavelets, Fractals, and Fourier Transforms, Farge M, Hant J, Vascillicos C, eds, Oxford UP, 1993,151-193

[8] Field DJ. What is the goal of sensory coding? *Neural Computation*, 1994, 6, 559-601.

[9] Foldiak P. Sparse coding in the primate cortex, In: The Handbook of Brain Theory and Neural Networks, Arbib MA, ed, MIT Press, 1995, 895-989

[10] Olshausen BA, Field DJ. Emergence of simple cell receptive field properties by learning a sparse code for natural images. *Nature*, 1996a, 381, 607- 609

[11] Olshausen BA, Field DJ. Natural image statistics and efficient coding. *Network*, 1996b, 7, 333-339.

[12] Ruderman DL.The statistics of natural images. Network : Computation in *Neural Systems*. 1994, 5, 517-548.

[13] Olshausen BA. Principles of Image Representation in Visual Cortex. In: Visual Neurosciences, Chalupa, LM & Werner, JS, eds, The MIT Press, Cambridge, Massachusetts, 2004,1603-1615.

[14] Sejnowski T. Time for a new neural code ? Nature, 1995, 376, 21-22.

[15] Aloso JM, Usrey WM, Reid RC. Precisely correlated firing in cell of the lateral geniculate nucleus. *Nature*, 1996, 383, 815-819.

[16] Mainen ZF, Sejnowski TJ. Reliability of spike timing in neocortical neurons. *Science*, 1995, 268, 1503-1506.

[17] Engel AK, Konig P, Kreiter AK, et al. Temporal coding in the visual cortex: new vistas on integration in the nervous systems. *Trends in Neuroscience*, 1992, 15(6): 218-226.

[18] Riehle A, Grum S, Diesman M, et al. Spike synchronization and rate modulation differentially involved in motor cortical function. *Science*, 1997, 278, 1959-1953.

[19] Hubel DH, Wiesel TN. Receptive fields, binocular interaction and functional architecture in the cat's striate cortex. *J. Physical*, 1962, 160, 106-154.

[20] Hubel DH, Wiesel TN. Receptive fields and functional architecture of monkeys striate cortex. *J. Physical*, 1968, 195, 215-243

[21] Ishai A , Ungerleider LG, Martin A, et al. Distributed representation of objects in the human ventral visual pathway. *Proc Natl Acad Sci USA*, 1999, 96, 9379-9384

[22] Grigorescu C, Petkov N, Westenberg MA. Contour detection by band-limited noise and its relation to non-classical receptive field inhibition. *IEEE Trans On Image Processing*, 2003, 12, 7: 729-739

[23] Lennie P. The cost of cortical computation. *Current Biology*, 2003, 38: 101-109

[24] Daugman JG. Complete discrete 2-D Gabor transforms by neural networks for image analysis and compression. *IEEE Transact. Acoustics, Speech Signal Process*, 1988, 37, 6, 1160-1179

[25] Lee TS. Image representation using 2D Gabor wavelets. *IEEE Trans Pattern Anal,* 1996, **18**, 959–971

[26] Jones JP, Palmer LA. The two-dimensional spatial structure of simple receptive fields in cat striate cortex. *J Neurophysiol*, 1987, 58, 1187–1211

[27] Riesenhuber M, Poggio T. Hierarchical models of object recognition in cortex. *Nat Neurosci*, 1999, 2, 11, 1019-1025.

[28] Nicholls JG, Martin AR, Wallace BG, et al. From Neuron to Brain. Fourth Edition, Sinauer Associates, Inc, 2001.

[29] Sugase Y, Yamane S, Ueno S, et al. Global and fine information coded by signal neurons in the temporal visual cortex. *Nature*, 1999, 400, 869-873.

[30] Young MP, Yamane S. Sparse population coding of faces in the inferotemporal cortex. *Science*, 1992, 256, 1327-1331

[31] Marr D. Vision: A computational investigation into the human representation and processing of visual information. New York: Freeman, 1982

[32] Zhao Songnian, Xiong Xiaoyun, Yao Guozheng, et al. A computational model as neurodecoder based on synchronous oscillation in the visual cortex. *Neural Comput*, 2003, 15, 2399-2418

[33] Ahissar E. Temporal-code to rate-code conversion by neuronal phase-looked loop. *Neural Comput*, 1998, 10, 579-650

[34] Ahissar E, Haidarliu S, Zacksenhouse M. Decoding temporally encoded sensory in put by cortical oscillations and thalamic phase comparators, *Proc Natl Acad Sci USA*, 1997, 94, 11633-11638

[35] Klapper J, Flankle JA. Phase-locked and frequency-feedback systems: Principles and techniques. Orlando, FL: Academic Press, 1972

[36] Kay SM. Fundamentals of statistical signal processing, Vol, I, Detection theory, Prentice Hall PTR, 1998, 520-550

[37] Snyder WE, Hairong Qi. Machine Vision. Cambridge: Cambridge University Press, 2004, 257-261

[38] Feichtinger HG, Strohmer T. Gabor Analysis and Algorithms: Theory and Application. Feichtinger HG, Strohmer T eds, Boston: Birkhaoser, 1998

[39] Issa NP, Rosenberg A, Husson TR Models and measurements of functional maps in V1. *J Neurophysiol*, 2008, 99, 2745-2754

[40] Rosa MGP. Visual maps in the adult primate cerebral cortex: some implication for brain development and evolution. *Braz J Med Biol Res*, 2002, 35, 12, 1485-1498

[41] Miikkulainen R, Bednar JA and Choe Y, et al. Computational Maps in the Visual Cortex. Berlin: Springer Science +Business Media, Inc, 2005

[42] Roelfsema PR. Cortical algorithms for perceptual grouping. *Annu Rev Neurosci*, 2006, 29, 203-27

[43] Adelson EH, Bergen JR. Spatiotemporal energy models for the perception of motion. *J Opt Soc Am A Opt Image Sci Vis*, 1985, 2, 284–299

[44] Baker TI, Issa NP. Cortical maps of separable tuning properties predict population responses to complex visual stimuli. *J Neurophysiol*, 2005, 94, 775–787

[45] Mante V, Carandini M. Mapping of stimulus energy in primary visual cortex. *J Neurophysiol*, 2005, 94, 788–798

[46] Simoncilli EP, Olshausen BO. Natural image statistics and neural representation. *Annu Rev Neurosci*, 2001, 24, 193-216

[47] O1shausen BA, Field DJ. Sparse coding with an over complete basis set: A strategy employed by V1? Visual Research, 1997, 37, 3311-3325

[48] Hyvarinen A, Oja E. A fast fixed-point algorithm for independent component analysis. *Neural Comput*, 1997, 9, 1483-1492

[49] Hyvarinen A, Hoyer PO. A two-layer sparse coding model learn simple and complex cell receptive fields and topography from natural images. *Vision Research*, 2002, 41, 18, 2413-2423

[50] Perrinet LU. Optimal signal representation in the neural spiking population codes: a model for the formation of simple cell receptive fields. (Report, Institute de Neuroscience Cognitives de la Mediterranee, 2008, CNRS/University of Provence, France)

[51] Zhao Songnian, Yao Li, Jin Zhen, et al. Sparse representation of global feature of visual image in human primary visual cortex: Evidence from fMRI. *Chinese Science Bulletin*, 2008, 53, 14, 2165-2174

[52] Carandini M, Demb JB, Mante V, et al. Do We Know What the Early Visual System Does? J Neuroscience, 2005, 25, 46, 10577–10597

[53] Olshausen BA, Field DJ. How close are we to understanding V1? *Neural Comput*, 2005, 17, 1665-1699

[54] Somers DC, Todorov EV, Siapas AG, et al. A local circuit integration approach to understanding visual cortical receptive fields. *Cerebral Cortex*, 1998, 8, 204-217.

[55] Troyer TW, Krukowski AE, Miller KD. "LGN Input to Simple Cells and Contrast-Invariant Orientation Tuning: An Analysis." *J Neurophysiol,* 2002, 87, 2741-2752.

[56] Swindale NV. Feedback decoding of spatially structured population activity in cortical maps. *Neural Comput*, 2007, 20, 1, 176-204.

[57] Larsson J, Landy MS, Heeger DJ. Orientation-selective adaptation to first- and second-order patterns in human visual cortex. *J Neurophysiol*, 2006, 95, 862–881.

[58] **Ringach** DL. Mapping receptive fields in primary visual cortex. *J.Physiol*, 2004, 558, 3, 717-728.

[59] Singh G, Memoli F, Ishkhanov T, et al. Topologocal analysis of population activity in visual cortex. Journal of Vision, 2008, 8, 11, 1-18

[60] Bednar JA, Miikkulainen R. Joint maps for orientation, eye, and direction preference in a self-organizing model of V1. *Neurocomputing*, 2006, 69, 1272–1276.

[61] Rybak IA, Gusakova VI, Golovan AV, et al. A model of attention-guided visual perception and recognition. *Vis. Research*, 1998, 38, 2387-2400.

In: Visual Cortex: Anatomy, Functions ... ISBN: 978-1-62100-948-1
Editors: J.M. Harris et al. pp. 165-184 © 2012 Nova Science Publishers, Inc.

Chapter 6

BIOLUMINESCENCE IMAGING OF *ARC* EXPRESSION DETECTS ACTIVITY-DEPENDENT AND PLASTIC CHANGES IN THE VISUAL CORTEX OF ADULT MICE

Hisashi Mori and Hironori Izumi
Department of Molecular Neuroscience,
Graduate School of Medicine and Pharmaceutical Sciences,
University of Toyama, Toyama 930-0194, Japan.

ABSTRACT

The activity-regulated cytoskeleton-associated protein gene *(Arc)* is one of the immediate early gene markers in the visual cortex up-regulated by light stimuli. The expression of *Arc* in the brain correlates with various sensory processes, natural behaviors, and pathological conditions. Arc is also involved in synaptic plasticity during development. Thus, *in vivo* monitoring of *Arc* expression is useful for the analysis of physiological and pathological conditions in the brain. We generated a novel transgenic mouse strain to monitor the neuronal-activity-dependent *Arc* expression using bioluminescence signals *in vivo*. Using the bacterial artificial chromosome (BAC) containing the entire mouse *Arc*, we introduced the firefly-derived luciferase (Luc) gene at the translational initiation site of *Arc* by homologous recombination in *Escherichia coli (E. coli.)* to generate the BAC transgene construct. We injected the construct into fertilized one-cell embryos and obtained transgenic mouse strains.

Immunohistochemical analysis revealed the strong coexpression of endogenous Arc and exogenous Luc in the neuronal soma in layers 4 and 6 in the visual cortex. Because of the very high sensitivity with a high signal-to-noise ratio, we successfully detected the changes in bioluminescence signal intensity in the visual cortex under the light and dark conditions. These changes correlated well with the changes in the expression levels of Arc and Luc examined by Western blot analysis. Visual deprivation by monocular eye enucleation (ME) resulted in the significant decrease in bioluminescence signal intensity in the contralateral posterior region within 24 hr. Interestingly, one month after ME, there was no significant difference in bioluminescence signal intensity between the right and left visual areas. These neuronal-activity-dependent plastic changes in the bioluminescence signal intensity in the mouse visual cortex after visual deprivation suggest structural plasticity after peripheral lesions in adults. Our novel mouse strain will be valuable for the continuous monitoring of neuronal-activity-dependent *Arc* expression in the visual cortex under physiological and pathological conditions.

1. INTRODUCTION

Recently, the mouse has been adopted by a number of researchers as a tool in the field of visual systems. The reasons for this adoption are in part due to the relative ease of modification of genetic information in this species. Complete genomic sequences of some inbred mouse strains are available (http://uswest.ensembl.org/Mus_musculus/Info/Index). The production of mice with precisely defined changes in gene sequence, namely, transgenic (Tg) and gene-knockout (KO) mice is based on the constructions of transgenic vectors using standard recombinant DNA technology and manipulation of long genomic DNA sequences (more than 200 kilo base pairs (kbps)) of mouse in bacterial artificial chromosome (BAC) with homologous recombination in *E. coli.* Many gene knockout mice were used for the analysis of the molecular basis of plasticity in the visual cortex [53]. Information on gene expression pattern in the mouse brain are also available in Allen Brain Atlas (http://www.brain-map.org/). Furthermore, there are other practical advantages to studying cortical plasticity in mice, such as their small size and short generation time, and various aspects of mouse visual system function and plasticity can be readily assessed using behavioral tests [41].

The mouse visual cortex is a part of the six-layer neocortex similar to the visual cortices of other mammals. Cytoarchitectural analyses define the areas that constitute the primary visual cortex (V1) in the mouse, namely, areas 17, 18a and 18b [5, 36, 48, 55]. Area 17 contains one complete representation of the contralateral visual hemifield with the zero vertical meridian represented close to the borders with area 18a (located laterally to area 17), and with gradually more peripheral vertical meridians represented more medially toward the border with area 18b [8, 24, 48, 55, 59]. The binocular zone of the mouse visual cortex occupies approximately the lateral one-third of area 17 and can be distinguished histologically from its higher acetylcholinesterase activity than the other areas [1]. Furthermore, a small (about 10°) part of the ipsilateral hemifield is represented in a small region of area 17 between the representation of the zero vertical meridian and the border with area 18a [8, 12, 55]. About 70% of cells recorded from the binocular segment of V1 respond to appropriate visual stimuli presented via either eye [8, 9, 31]. Even though mice do not have anatomically distinct ocular dominance (OD) columns, they have a critical period during their early postnatal life in which the relative representations of the two eyes in the binocular region of the visual cortex are sensitive to monocular deprivation (MD) [1, 9, 12]. The temporary closure of one eye for MD results in an overall strengthening of the open-eye representation in the visual cortex. Various techniques have been developed to determine OD in the mouse visual cortex by single unit recording [9, 12, 20], recording of visually evoked potentials [11, 22, 39, 46], optical imaging [3, 18, 21], calcium imaging with two-photon microscopy [34, 43, 50], and gene expression analysis of activity-regulated genes, such as *c-fos* [37] or *Arc* [51], in histological sections.

In this review, we summarize our recently reported findings of studies using our novel BAC Tg mouse strain to monitor the neuronal-activity-dependent *Arc-Luc* expression in the mouse visual cortex using bioluminescence signals [23] and compare our mouse strain with other *Arc* reporter mouse strains.

2. Neuronal Immediate-Early Gene *Arc*

The activity-regulated cytoskeleton-associated protein gene (*Arc*, also known as *Arg3.1*) was first identified as an immediate early gene induced by seizure and depolarization in hippocampal neurons [27, 28]. Hippocampal *Arc* transcripts in rodents are also induced during exploration of a novel environment [14, 17, 54], and the levels of *Arc* expression correlate with learning in hippocampus-dependent spatial learning tasks [16]. Furthermore, the decrease in the expression level of Arc following infusion of antisense oligonucleotides into the rat hippocampus [15] or the targeted deletion of *Arc* in the mouse [38] interferes with hippocampus-dependent synaptic plasticity and learning and memory. One of the proposed roles of Arc in the regulation of synaptic plasticity is its involvement in α-amino-3-hydroxy-5-methyl-4-isoxazole propionic acid (AMPA)-type glutamate receptor endocytosis [7, 44, 47]. Regulatory mechanisms of *Arc* expression are extensively studied [35].

In the visual cortex, *Arc* expression is induced by light exposure and is used for the mapping of neuronal activity [25, 51]. A study using *Arc* knockout (KO) mice revealed the functions of Arc in the orientation specificity of visual cortical neurons [57]. Furthermore, Arc is suggested to be required in experience-dependent processes that normally establish and modify synaptic connections in the visual cortex [30].

From these studies, *in vivo* monitoring of Arc expression is useful for the analysis of physiological conditions in the brain. However, most of the methods for Arc detection in the above-mentioned studies are *in situ* hybridization and immunohistochemical analysis applied to fixed brain sections. Recently, three research groups have reported the successful *in vivo* imaging of *Arc* expression under physiological and pathological conditions using fluorescent protein gene probes. These probes are the d2-type enhanced green fluorescent protein (EGFP) gene (*d2EGFP*) knocked into the *Arc* locus in mice [57], the fluorescent protein reporter gene *Venus* linked to the 7.1 kbp promoter region of *Arc* in conventional transgenic (Tg) mice [10], and the short-life form of *EGFP, d4EGFP,* introduced into BAC containing *Arc* in BAC Tg mice [13]. Wang et al. [57] successfully monitored Arc expression at the single neuron level from cortical layers II to IV during visual cortical activation using visual stimuli. The two research groups obtained superficial brain fluorescence signals of Venus and d4EGFP from the visual cortex and somatosensory cortex [10, 13] with some autofluorescence noise. These groups could not detect changes in signal intensity associated with experience-dependent plasticity.

3. Generation of Novel Mouse Strain to Detect Expression of *Arc*

One of the powerful *in vivo* imaging probes is luciferase (Luc). Bioluminescence signals emitted from Luc provide a much higher sensitivity than fluorescence signals for noninvasive detection *in vivo* [2] and are nontoxic, allowing continuous and repeated recording over a long period [29]. Moreover, bioluminescence signals can be detected from the depths of the brain with a very high signal-to-noise (S/N) ratio [40, 45]. These properties of bioluminescence imaging are suitable for the neuroimaging of contextually relevant spatiotemporal expression of Arc in mice. Recently, we have generated and reported on a novel BAC Tg mouse strain to monitor the neuronal-activity-dependent *Arc-Luc* expression using bioluminescence signals [23]. Using the *Arc-Luc* Tg mice, we successfully detected massive plastic changes in bioluminescence signal intensity in the adult mouse visual cortex after visual deprivation.

To generate the Tg mouse strain, we first obtained the mouse BAC clone RP24-388I10 carrying a 221-kbp insert containing the entire *Arc* from the BACPAC Resources Center CHORI. The BAC clone contained the entire *Arc* and about 83 kbp of the 5' upstream sequence and about 125 kbp of the 3' downstream sequence. We introduced *Luc* at the translational initiation site of *Arc* by homologous recombination in *E. coli.* to generate the BAC transgene construct pTg-Arc-Luc (Figure 1).

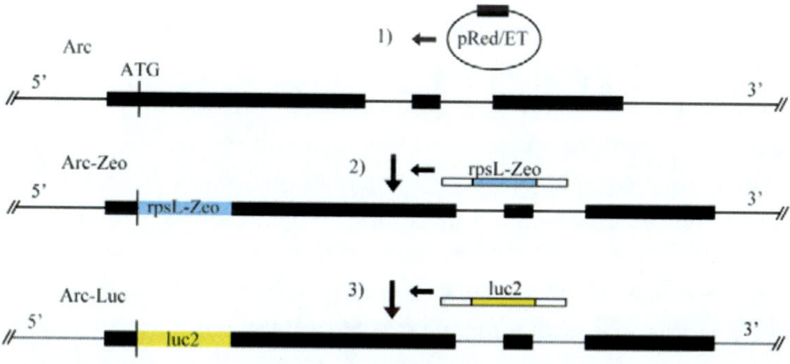

Figure 1. Generation of *Arc-Luc* BAC transgene by homologous recombination in *E. coli.* Steps of homologous recombination in *E. coli* to generate Arc-Luc BAC transgene. First, the Red/ET recombination proteins expression plasmid is introduced into *E. coli* containing Arc BAC DNA. Second, the PCR fragment containing the drug selection cassette (rspL-Zeo) attached with ~50 bps homologous regions at each end is introduced. Third, the DNA fragment containing the luciferase coding region (Luc) attached with about 300 bps homologous regions of Arc at each end is transformed. These steps of homologous recombination are confirmed on the basis of the resistance of *E. coli* to antibiotics, PCR, and Southern blot analysis results, and DNA sequencing.

After the purification of pTg-Arc-Luc DNA, the transgene was linearized by *Not*I digestion, and purified BAC DNA was microinjected into pronuclei of fertilized one-cell embryos from C57BL/6 mice.

Among 49 candidates, we identified 10 lines of mice carrying the *Arc-Luc* transgene by PCR and Southern blot analyses of genomic DNA prepared from tail biopsy specimens. The integrated *Arc-Luc* transgene was stably transmitted to the next generation and one of the lines with strong bioluminescence signals was used for the bioluminescence imaging study.

4. Bioluminescence Imaging of Arc Expression in the Visual Cortex

The Tg mice were anesthetized by sodium pentobarbital injection or inhalation of isoflurane before and during imaging. Because the black fur of the C57BL/6 strain attenuates photon emission, the fur on the head of mice was shaved. The mice were injected with luciferin, a substrate of luciferase under anesthesia. Ten minutes after the luciferin injection, the bioluminescence signal intensity in *Arc-Luc* Tg mice was measured using an *in vivo* imaging system consisting of a dark chamber and a cooled charge-coupled device (CCD) camera. Bioluminescence images were taken for 30 or 180 sec with 4 x 4 binning without using an optical filter. Pseudocolored luminescent images representing the spatial distribution of emitted photons were overlaid on photographs of mice taken in the chamber under a dim light.

To detect light-induced bioluminescence signal intensity changes in the visual cortex, the *Arc-Luc* mice were anesthetized with sodium pentobarbital, and bioluminescence signal intensity was measured for 180 sec. For quantitative analysis, we set a template image of the bioluminescence signals of the brain of each mouse placed under light condition, then the regions of interest (ROIs), including the somatosensory and visual cortex areas, were selected as follows. Using the bioluminescence signal image of the cerebral hemisphere, we defined the long axis and formed a circle with its center in the middle of the long axis and with a radius of 2 mm for the region containing the somatosensory cortex. Similarly, we formed another circle with a radius of 2 mm and with its center located at 1/4 of the length of the long axis from the edge of the caudal border of a bioluminescence signal image to define the ROI containing the visual cortex, as shown in Figure 2. The photon counts in the central regions containing the somatosensory cortex and posterior regions containing the visual cortex ranged from 8219 to 24044 and from 7080 to 20953, respectively. The photon counts in the central regions and posterior regions of Tg mice under normal light condition were highly correlated (γ = 0.989, n = 32).

In all the experiments, background bioluminescence signal intensity was measured in wild-type (WT) mice injected with luciferin and the *Arc-Luc* Tg mice injected with phosphate-buffered saline (PBS) instead of luciferin. The obtained background bioluminescence signal intensity was subtracted from measured bioluminescence signal intensity. Data were expressed as the mean number of counted photons in the ROI. Bioluminescence signal intensity was calculated from bioluminescence images by ROI analysis using NIH ImageJ.

Using a cooled CCD camera, we detected bioluminescence signals in the nose and head regions of the Tg mice in a dark chamber (Figure 2). Strong signals were detected in the cerebral cortical areas of the brain. To determine the source of the detected bioluminescence signals, we prepared coronal brain slices from the *Arc-Luc* Tg mice and incubated them with luciferin. We detected bioluminescence signals in the cerebral cortex and hippocampus in the brain slices. We were unable to detect any significant bioluminescence signal in other brain regions. These distribution patterns of bioluminescence signals were similar to those reported for *Arc* mRNA in the mouse brain [51].

We further examined the localization of endogenous Arc and exogenous Luc proteins in the *Arc-Luc* Tg mouse brain by double immunofluorescence staining. Strong signals of Arc were detected in the neuronal soma in layers 4 and 6 in the cortex (Figure 3) and these distribution patterns were consistent with those reported for Arc [28, 42] and its mRNA [51]. The immunofluorescence signals of the exogenous Luc were also detected in the neurons in layers 4 and 6 in the visual cortex (Figure 3).

Most of the immunofluorescence signals of Arc and Luc merged. From these findings, we concluded that the bioluminescence light emitted from the Tg Luc protein mimicked the expression pattern of *Arc* in the brain.

Arc is one of the neuronal immediate early gene markers in the visual cortex up-regulated by light stimuli [51]. To quantify the emitted photons from ROIs, we set a template image of the bioluminescence signals of the brain of mice under light conditions.

The ROIs including the somatosensory and visual cortex areas were selected, as shown in Figure 2, and the photon emission intensities of these regions were measured as described above.

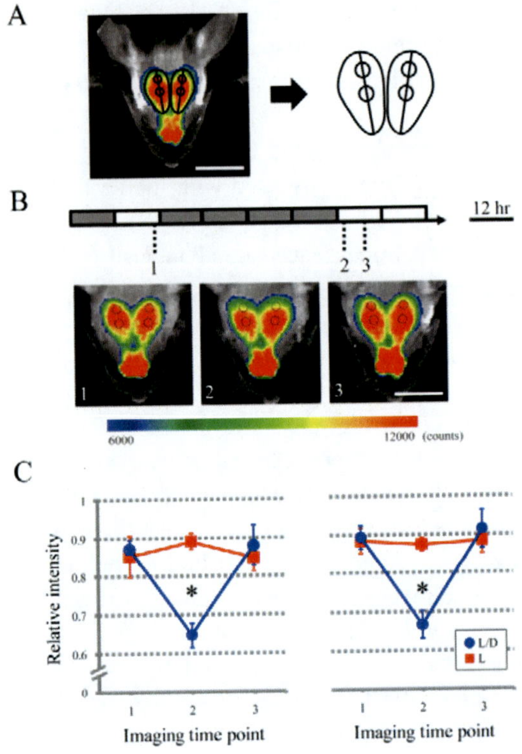

Figure 2. Changes in bioluminescence signal intensity in visual cortex of *Arc-Luc* Tg mice under light and dark conditions. A Definition of regions of interest (ROIs). A template image of bioluminescence signals of the brain under light condition was set and the ROIs including the somatosensory and visual cortex areas (circles) were identified. Then, the bioluminescence signal intensities of these regions were measured as described in the text. Scale bar, 10 mm. B Protocol for light and dark conditions and imaging (upper panel) and obtained bioluminescence images (lower panels; pseudocolored, 6000-12000 counts). The dark (gray box, 12 hr) and light (open box, 12 hr) conditions and the time points of imaging (1-3) are indicated. Imaged areas of the visual and somatosensory cortices are indicated by dotted circles. Scale bar, 10 mm. C Changes in relative bioluminescence signal intensity in visual cortex of *Arc-Luc* Tg mice (n = 14) under light (L) and dark (D) conditions (L/D). The relative intensities of bioluminescence signals in the left (left panel) and right (right panel) visual cortex regions were calculated from the measured bioluminescence signal intensity normalized with that of the somatosensory cortex. The control group of mice (n = 4) was placed under continuous light (L) condition, and bioluminescence signal intensity was measured at the same time points as those groups under L and D conditions. The data represent mean ± SD. *$p < 0.05$; two-tailed Student's *t*-test. (Modified from ref. 23 with permission).

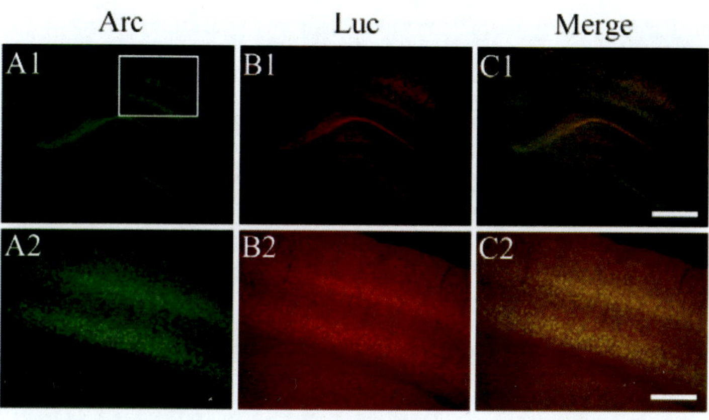

Figure 3. Distribution patterns of Arc and luciferase in visual cortex. A Immunohistochemical staining of slices with anti-Arc antibody (green, A1). Rectangles for higher-magnification images in the visual cortex (A2) are indicated in A1. B Immunofluorescence staining of the slices with anti-luciferase antibody (red, B1). Higher-magnification images of immunofluorescence signals in the visual cortex (B2). C Merged images of immunofluorescence signals of slices stained with anti-Arc and anti-luciferase antibodies (yellow, C1). Higher-magnification images of immunofluorescence signals in visual cortex (C2). Scale bar, 750 μm (A1, B1, and C1) and 300 μm (A2, B2, and C2). (Modified from ref. 23 with permission).

To examine the effect of light stimuli on the bioluminescence signals in the *Arc-Luc* Tg mice, we placed the Tg mice under dark condition for 48 hr. After this dark condition, the bioluminescence signal intensity of the posterior ROIs of the cerebral cortex apparently decreased (Figure 2).

Seven hr of continuous light exposure resulted in the recovery of the bioluminescence signal intensity in these regions (Figure 2). These regions contain the visual cortex of the mouse [8]. To evaluate quantitatively the changes in the bioluminescence signal intensity in these regions, we calculated the relative bioluminescence signal intensity in these regions in comparison with that in the central regions of the cerebral cortex containing the reported somatosensory areas of the mouse brain [4]. As shown in Figure 2, placing the mice under dark condition for 48 hr significantly decreased the bioluminescence signal intensity in the bilateral regions containing the visual cortex, which recovered after 7 hr of light exposure. Under continuous light condition, no change in relative bioluminescence signal intensity was detected (Figure 2). Furthermore, these changes in bioluminescence signal intensity correlated well with the changes in the expression levels of Arc and Luc determined by Western blot analysis. From these findings, we assume that the

Arc-Luc Tg mice are valuable for monitoring the neuronal activity in the visual cortex induced by light.

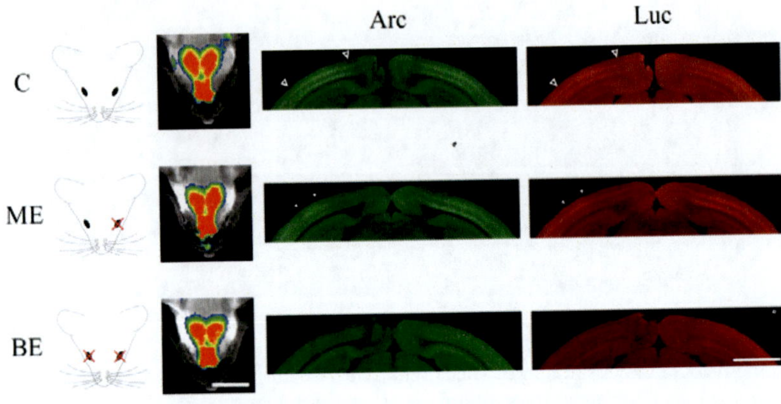

Figure 4. Changes in bioluminescence signal intensity in visual cortex of *Arc-Luc* Tg mice after visual deprivation. Illustrated figures, bioluminescence images, immunostaining of Arc (green), and immunostaining of Luc (red) of visual deprivation by eye enucleation in mice. Bioluminescence images (pseudocolored, 6000-12000 counts) were obtained over the head region of control mice before enucleation (C, upper) and after monocular enucleation (ME, middle) and binocular enucleation (BE, lower). Immunopositive signal patterns of Arc and luciferase in the visual cortex were similar to the obtained bioluminescence signal patterns in the visual cortex shown. The Arc- and Luc-positive visual cortical region is indicated by large arrowheads. The binocular visual zone is indicated by small arrowheads in ME. Scale bars, 10 mm in bioluminescence images; 2 mm in brain sections. (Modified from ref. 23 with permission).

We further confirmed the possibility that the bioluminescence signals detected in posterior brain regions correspond to the Luc expression induced by visual stimuli. The mouse visual cortex contains a large (~70%) monocular zone that receives inputs only from the contralateral eye and a small (~30%) binocular zone that receives inputs from both eyes [1]. Visual deprivation by monocular eye enucleation (ME) resulted in the significant decrease in bioluminescence signal intensity in the contralateral posterior brain region within 24 hr (Figure 4). Visual deprivation by binocular eye enucleation (BE) further decreased the bioluminescence signal intensity in the contralateral posterior brain region (Figure 4). These decreases in bioluminescence signal intensity in the posterior brain region correlated with the decreased expression levels of Arc and Luc, as revealed by immunofluorescence analyses (Figure 4). From these findings, we were able to monitor the visual-activity-dependent changes in the bioluminescence signal intensity in the visual cortex in the *Arc-Luc* Tg mouse brain.

5. Detection of Plastic Changes in *Arc* Expression in the Visual Cortex of Adult Mice

Using *Arc-Luc* Tg mice, we detected the plastic changes in bioluminescence signal intensity after the visual deprivation in adult mice. After the ME at the age of 16 weeks, we measured the bioluminescence signal intensity in the mouse brain for three consecutive months (Figure 5). Four days after the ME, the bioluminescence signal intensity in the right visual area corresponding to the cortex contralateral to the enucleated eye decreased significantly (Figure 5).

However, one month after ME, there was no significant difference in bioluminescence signal intensity between the right and left visual areas (Figure 5). The recovered intensity of bioluminescence signals from the visual cortex contralateral to the enucleated eye was maintained for another two months. There was no obvious change in relative bioluminescence signal intensity in the ipsilateral visual cortex after ME (Figure 5).

The intensity of the bioluminescence signals obtained from both visual cortices changed equally depending on the lighting condition at latest two months after ME. These findings clearly indicate the plastic changes in *Arc-Luc* expression level in the visual cortex of the adult mouse after continuous ME.

The immediate early gene *Arc* is induced in the brain by various stimuli and is involved in synaptic plasticity during development as revealed by KO mouse studies [30, 38, 57]. Arc is considered to be involved in several forms of experience-dependent synaptic plasticity such as long-term potentiation [15], long-term depression [44], and homeostatic synaptic plasticity [47]. In the visual cortex, endogenous *Arc* expression is induced mainly in layer 4 neurons by light stimuli and is dependent on the activation of *N*-methyl-D-aspartate (NMDA) receptor channels [27, 28, 49]. The mouse visual cortex is divided into the monocular and binocular zones; in the binocular zone, the competition between the activities from inputs from both eyes continuously occurs [12].

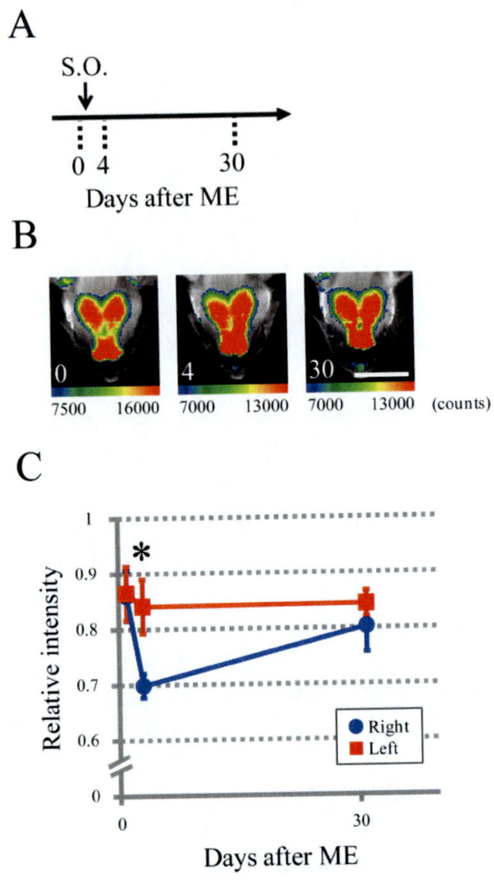

Figure 5. Detection of plastic changes in bioluminescence signal intensity in visual cortex after monocular deprivation in adult *Arc-Luc* Tg mice. A Protocol for surgical operation (S.O.) and bioluminescence imaging 0, 4, and 30 days after ME in *Arc-Luc* Tg mice. B Representative images of bioluminescence (pseudocolored, 7000-16000 counts) in visual cortex of *Arc-Luc* Tg mice taken at time points shown in A. Scale Bar, 10 mm. C Relative intensity of bioluminescence signals obtained at indicated time points shown in A. The relative intensity of bioluminescence signals in the visual cortex region is calculated in comparison with that of the somatosensory cortex. The data represent mean ± SD (n = 3-9). *$p < 0.05$; two-tailed Student's *t*-test. (Modified from ref. 23 with permission).

We detected a decrease in the bioluminescence signal intensity of Arc-Luc in the visual cortex contralateral to the enucleated eye four days after the surgery and the recovery of the signal intensity in the same area within one month. The recovery of bioluminescence signal intensity depended on the lighting condition, at the latest two months after ME.

These findings suggest that the recovery of bioluminescence signal intensity in this visual cortex depends on the neuronal activity in the normal eye. Adult mice have a greater potential for experience-dependent plasticity than previously considered, because the MD of the dominant contralateral eye leads to a persistent, NMDA-receptor-dependent enhancement of weak ipsilateral-eye inputs [46]. Keck et al. [26] monitored by intrinsic-signal detection and two-photon imaging the functional and structural alterations in the adult mouse visual cortex after focal retinal lesioning. They suggested the activity-dependent establishment of new cortical circuits that leads to the recovery of visual responses.

On the other hand, Mrsic-Flogel et al. [32] have monitored by two-photon calcium imaging the short-term homeostatic feedback up-regulation of neuronal activity in the visual cortex after monocular deprivation. Furthermore, in the binocular zone of the adult mouse visual cortex, the extent of *Arc* induction after stimulation of the ipsilateral nondeprived eye increases and is expanded four days after ME [51]. In our study, we were unable to detect the early homeostatic up-regulation of Arc-Luc bioluminescence signals in the visual cortex four days after ME. Thus, the involvement of a homeostatic feedback process in the recovery of bioluminescence signal intensity observed one month after ME might be small. Rather, competition with neuronal-activity-dependent synaptic plasticity may be involved in the recovery process. To examine the mechanism underlying the recovery of bioluminescence signal intensity in the visual cortex contralateral to the enucleated eye, further histological examination of changes in axonal branching in the visual cortex in our ME mice is necessary.

6. Comparisons of Arc Imaging Methods

To precisely reproduce the *in vivo* expression pattern of *Arc*, we generated BAC Tg mouse strains. By homologous recombination in *E. coli* [33], we were able to precisely modify the BAC DNA containing *Arc* to insert *Luc* easily and quickly (Figure 1). The expression of large DNA transgenes such as BAC vectors can accurately reflect the transcription pattern of an endogenous chromosomal gene in a dose-dependent and integration-site-independent manner [19]. The findings of immunohistochemical and Western blot analyses using the anti-Arc and anti-luciferase antibodies supported the spatiotemporal expression patterns of the bioluminescence signals accurately reflecting the endogenous Arc and transgenic Luc protein expression patterns. Recently, three research groups have reported the successful *in vivo* imaging of *Arc* expression by the knockin of *d2EGFP* into the *Arc* locus [57], the conventional transgenic method using the 7.1-kbp 5' upstream region of *Arc* and the fluorescent protein reporter *Venus* [10], and the BAC transgenic method using the short-life form *d4EGFP* [13] (Table 1). The reported expression patterns of *Arc-dVenus* in the visual cortex of the transgenic mice were similar to those obtained by us [23] and others [25, 51].

Table 1. Comparison of Arc-reporter mouse lines

Mouse line	Method	Promoter	Reporter	Ref.
Arc-GFP	Knockin	Endogenous	d2EGFP	57
Arc-dVenus	Conventional Tg	7.1-kb upstream	dVenus	10
TgArc/Arg3.1-d4EGFP	BAC Tg	BAC	d4EGFP	13
Arc-Luc BAC Tg	BAC Tg	BAC	Luc	23

In contrast to the superficially obtained fluorescence signals using Venus and d4EGFP [10, 13] with some autofluorescence, our results suggest that we were able to detect bioluminescence signals from the depths of the brain such as the hippocampus with a very high S/N ratio. Furthermore, bioluminescence imaging is the most sensitive method of small-animal imaging [45], and no external or cytotoxic excitation light is required to generate bioluminescence signals. We were able to perform long-term imaging of Arc-Luc expression for more than 3 months noninvasively. These features of bioluminescence imaging are suitable for the continuous monitoring and contextually relevant spatiotemporal expression of Arc in mice. In the case of our *Arc-Luc* Tg mouse strain, there are also some limitations in comparison with other *Arc*-reporter mice. Using our *Arc-Luc* Tg mouse strain, we were unable to analyze the *Arc-Luc* expression at a single-cell level *in vivo* as previously reported [13, 57]. On the other hand, a bioluminescence image could be overlaid on a photograph of a mouse used as an anatomical reference, as we showed here; however, these 2D images lack 3D information. Because algorithms have been developed to assign the position of the light source more precisely [6, 58], further analysis will reveal the source of the bioluminescence signals in the brain. In our Arc-Luc Tg mice, the temporal expression patterns of the obtained bioluminescence signals and reported Arc protein may be slightly different, because we used wild-type firefly luciferase with a half-life of about 3 hr [52] in contrast to the endogenous Arc protein with a half-life of the about 2 hr [56, 57]. Nevertheless, despite these limitations of our mouse strain, we were able to easily monitor the neuronal-activity-dependent *Arc-Luc* expression and detected the plastic changes in *Arc* expression in the brain.

Conclusions

Bioluminescence-based imaging is a powerful method for monitoring neuronal activity in the brain. Here, we successfully generated a novel BAC Tg mouse strain, *Arc-Luc,* for monitoring the neuronal-activity-dependent *Arc* expression using bioluminescence signals in the mouse brain. Changes in bioluminescence signal intensity in the *Arc-Luc* mouse visual cortex were induced by light and dark conditions and eye enucleation.

The intensity of these bioluminescence signals correlated with endogenous Arc and exogenous Luc protein expression levels. Interestingly, we detected the recovery of bioluminescence signal intensity in the visual cortex one month after ME, suggesting the usefulness of our mouse strain for the detection of plastic changes in neuronal-activity-dependent Arc expression in the adult brain.

Acknowledgements

We thank Professor Kaoru Inokuchi for help in fluorescence microscopy and Dr. Tetsuya Ishimoto for help in bioluminescence imaging. This work was supported by a grant from the Food Safety Commission, Japan (No. 1001).

References

[1] Antonini, A., Fagiolini, M., Stryker, M.P. (1999). Anatomical correlates of functional plasticity in mouse visual cortex. *J. Neurosci., 19,* 4388-4406.

[2] Brandes, C., Plautz, J.D., Stanewsky, R., Jamison, C.F., Straume, M., Wood, K.V., Kay, S.A., Hall, J.C. (1996). Novel features of drosophila

period transcription revealed by real-time luciferase reporting. *Neuron, 16,* 687-692.

[3] Cang, J., Kalatsky, V.A., Lowel, S., Stryker, M.P. (2005). Optical imaging of the intrinsic signal as a measure of cortical plasticity in the mouse. *Vis. Neurosci., 22,* 685-691.

[4] Caviness, V.S., Jr. (1975). Architectonic map of neocortex of the normal mouse. *J. Comp. Neurol.*, 164, 247-263.

[5] Caviness, V.S., Jr., and Frost, D.O. (1980). Tangential organization of thalamic projections to the neocortex in the mouse. *J. Comp. Neurol., 194,* 335-367.

[6] Chaudhari, A.J., Darvas, F., Bading, J.R., Moats, R.A., Conti, P.S., Smith, D.J., Cherry, S.R., Leahy, R.M. (2005). Hyperspectral and multispectral bioluminescence optical tomography for small animal imaging. *Phys. Med. Biol., 50,* 5421-5441.

[7] Chowdhury, S., Shepherd, J.D., Okuno, H., Lyford, G., Petralia, R.S., Plath, N., Kuhl, D., Huganir, R.L., Worley, P.F. (2006). Arc/Arg3.1 interacts with the endocytic machinery to regulate AMPA receptor trafficking. *Neuron, 52,* 445-459.

[8] Dräger, U.C. (1975). Receptive fields of single cells and topography in mouse visual cortex. *J. Comp. Neurol., 160,* 269-290.

[9] Dräger, U.C. (1978). Observations on monocular deprivation in mice. *J. Neurophysiol., 41,* 28-42.

[10] Eguchi, M., and Yamaguchi, S. (2009). In vivo and in vitro visualization of gene expression dynamics over extensive areas of the brain. *NeuroImage, 44,* 1274-1283.

[11] Frenkel, M.Y., and Bear, M.F. (2004). How monocular deprivation shifts ocular dominance in visual cortex of young mice. *Neuron, 44,* 917-923.

[12] Gordon, J.A., and Stryker, M.P. (1996). Experience-dependent plasticity of binocular responses in the primary visual cortex of the mouse. *J. Neurosci., 16,* 3274-3286.

[13] Grinevich, V., Kolleker, A., Eliava, M., Takada, N., Takuma, H., Fukazawa, Y., Shigemoto, R., Kuhl, D., Waters, J., Seeburg, P.H., Osten, P. (2009). Fluorescent Arc/Arg3.1 indicator mice: a versatile tool to study brain activity changes in vitro and in vivo. *J. Neurosci. Methods,* 184, 25-36.

[14] Guzowski, J.F., McNaughton, B.L., Barnes, C.A., Worley, P.F. (1999). Environment-specific expression of the immediate-early gene Arc in hippocampal neuronal ensembles. *Nat. Neurosci., 2,* 1120-1124.

[15] Guzowski, J.F., Lyford, G.L., Stevenson, G.D., Houston, F.P., McGaugh, J.L., Worley, P.F., Barnes, C.A. (2000). Inhibition of activity-dependent arc protein expression in the rat hippocampus impairs the maintenance of long-term potentiation and the consolidation of long-term memory. *J. Neurosci., 20,* 3993-4001.

[16] Guzowski, J.F., Setlow, B., Wagner, E.K., McGaugh, J.L. (2001). Experience-dependent gene expression in the rat hippocampus after spatial learning: a comparison of the immediate-early genes Arc, c-fos, and zif268. *J. Neurosci., 21,* 5089-5098.

[17] Guzowski, J.F., Knierim, J.J., Moser, E.I. (2004). Ensemble dynamics of hippocampal regions CA3 and CA1. *Neuron,* 44, 581-584.

[18] Heimel, J.A., Hartman, R.J., Hermans, J.M., Levelt, C.N. (2007). Screening mouse vision with intrinsic signal optical imaging. *Eur. J. Neurosci., 25,* 795-804.

[19] Heintz, N. (2001). BAC to the future: the use of bac transgenic mice for neuroscience research. *Nat. Rev. Neurosci., 2,* 861-870.

[20] Hensch, T.K., Fagiolini, M., Mataga, N., Stryker, M.P., Baekkeskov, S., Kash, S.F. (1998). Local GABA circuit control of experience-dependent plasticity in developing visual cortex. *Science, 282,* 1504-1508.

[21] Hofer, S.B., Mrsic-Flogel, T.D., Bonhoeffer, T., Hubener, M. (2006). Prior experience enhances plasticity in adult visual cortex. *Nat. Neurosci., 9,* 127-132.

[22] Huang, Z.J., Kirkwood, A., Pizzorusso, T., Porciatti, V., Morales, B., Bear, M.F., Maffei, L., Tonegawa, S. (1999). BDNF regulates the maturation of inhibition and the critical period of plasticity in mouse visual cortex. *Cell, 98,* 739-755.

[23] Izumi, H., Ishimoto, T., Yamamoto, H., Nishijo, H., Mori, H. (2011). Bioluminescence imaging of Arc expression enables detection of activity-dependent and plastic changes in the visual cortex of adult mice. *Brain Struct. Funct., 216,* 91-104.

[24] Kalatsky, V.A., and Stryker, M.P. (2003). New paradigm for optical imaging: temporally encoded maps of intrinsic signal. *Neuron, 38,* 529-545.

[25] Kawashima, T., Okuno, H., Nonaka, M., Adachi-Morishima, A., Kyo, N., Okamura, M., Takemoto-Kimura, S., Worley, P.F., Bito, H. (2009). Synaptic activity-responsive element in the Arc/Arg3.1 promoter essential for synapse-to-nucleus signaling in activated neurons. *Proc. Natl. Acad. Sci. USA., 106,* 316-321.

[26] Keck, T., Mrsic-Flogel, T.D., Vaz Afonso, M., Eysel, U.T., Bonhoeffer, T., Hübener, M. (2008). Massive restructuring of neuronal circuits during functional reorganization of adult visual cortex. *Nat. Neurosci., 11,* 1162-1167.

[27] Link, W., Konietzko, U., Kauselmann, G., Krug, M., Schwanke, B., Frey, U., Kuhl, D. (1995). Somatodendritic expression of an immediate early gene is regulated by synaptic activity. *Proc. Natl. Acad. Sci. USA., 92,* 5734-5738.

[28] Lyford, G.L., Yamagata, K., Kaufmann, W.E., Barnes, C.A., Sanders, L.K., Copeland, N.G., Gilbert, D.J., Jenkins, N.A., Lanahan, A.A., Worley, P.F. (1995). Arc, a growth factor and activity-regulated gene, encodes a novel cytoskeleton-associated protein that is enriched in neuronal dendrites. *Neuron, 14,* 433-445.

[29] Martin, J.R. (2008). In vivo brain imaging: Fluorescence or bioluminescence, which to choose? *J. Neurogenet., 22,* 285-307.

[30] McCurry, C.L., Shepherd, J.D., Tropea, D., Wang, K.H., Bear, M.F., Sur, M. (2010). Loss of Arc renders the visual cortex impervious to the effects of sensory experience or deprivation. *Nat. Neurosci., 13,* 450-457.

[31] Metin, C., Godement, P., Imbert, M. (1988). The primary visual cortex in the mouse: receptive field properties and functional organization. *Exp. Brain Res.,* 69, 594-612.

[32] Mrsic-Flogel, T.D., Hofer, S.B., Ohki, K., Reid, R.C., Bonhoeffer, T., Hübener, M. (2007). Homeostatic regulation of eye-specific responses in visual cortex during ocular dominance plasticity. *Neuron, 54,* 961-972.

[33] Muyrers, J.P., Zhang, Y., Testa, G., Stewart, A.F. (1999). Rapid modification of bacterial artificial chromosome by ET-recombination. *Nucleic Acids Res., 27,* 1555-1557.

[34] Ohki, K., Chung, S., Ch'ng, Y.H., Kara, P., Reid, R.C. (2005). Functional imaging with cellular resolution reveals precise micro-architecture in visual cortex. *Nature, 433,* 597-603.

[35] Okuno, H. (2011). Regulation and function of immediate-early genes in the brain: beyond neuronal activity markers. *Neurosci. Res., 69,* 175-186.

[36] Olavarria, J. and Montero, V.M. (1989). Organization of visual cortex in the mouse revealed by correlating callosal and striate-extrastriate connections. *Vis. Neurosci., 3,* 59-69.

[37] Pham, T.A., Graham, S.J., Suzuki, S., Barco, A., Kandel, E.R., Gordon, B., Lickey, M.E. (2004). A semi-persistent adult ocular dominance

plasticity in visual cortex is stabilized by activated CREB. *Learn. Mem., 11,* 738-747.

[38] Plath, N., Ohana, O., Dammermann, B., Errington, M.L., Schmitz, D., Gross, C., Mao, X., Engelsberg, A., Mahlke, C., Welzl, H., Kobalz, U., Stawrakakis, A., Fernandez, E., Waltereit, R., Bick-Sander, A., Therstappen, E., Cooke, S.F., Blanquet, V., Wurst, W., Salmen, B., Bösl, M.R., Lipp, H.P., Grant, S.G., Bliss, T.V., Wolfer, D.P., Kuhl, D. (2006). Arc/Arg3.1 is essential for the consolidation of synaptic plasticity and memories. *Neuron, 52,* 437-444.

[39] Porciatti, V., Pizzorusso, T., Maffei, L. (1999). The visual physiology of the wild type mouse determined with pattern VEPs. *Vision Res., 39,* 3071-3081.

[40] Prescher, J.A., and Contag, C.H. (2010). Guided by the light: visualizing biomolecular processes in living animals with bioluminescence. *Curr. Opin. Chem. Biol.,* 14, 80-89.

[41] Prusky, G.T., West, P.W., Douglas, R.M. (2000). Behavioral assessment of visual acuity in mice and rats. *Vision Res., 40,* 2201-2209.

[42] Ramirez-Amaya, V., Vazdarjanova, A., Mikhael, D., Rosi, S., Worley, P.F., Barnes, C.A. (2005). Spatial exploration-induced Arc mRNA and protein expression: evidence for selective, network-specific reactivation. *J. Neurosci., 25,* 1761-1768.

[43] Regehr, W.G., and Tank, D.W. (1991). Selective fura-2 loading of presynaptic terminals and nerve cell processes by local perfusion in mammalian brain slice. *J. Neurosci. Methods, 37,* 111-119.

[44] Rial Verde, E.M., Lee-Osbourne, J., Worley, P.F., Malinow, R., Cline, H.T. (2006). Increased expression of the immediate-early gene arc/arg3.1 reduces AMPA receptor-mediated synaptic transmission. *Neuron, 52,* 461-474.

[45] Rice, B.W., Cable, M.D., Nelson, M.B. (2001). In vivo imaging of light-emitting probes. *J. Biomed. Opt.,* 6, 432-440.

[46] Sawtell, N.B., Frenkel, M.Y., Philpot, B.D., Nakazawa, K., Tonegawa, S., Bear, M.F. (2003). NMDA receptor-dependent ocular dominance plasticity in adult visual cortex. *Neuron, 38,* 977-985.

[47] Shepherd, J.D., Rumbaugh, G., Wu, J., Chowdhury, S., Plath, N., Kuhl, D., Huganir, R.L., Worley, P.F. (2006). Arc/Arg3.1 mediates homeostatic synaptic scaling of AMPA receptors. *Neuron, 52,* 475-484.

[48] Simmons, P.A., Lemmon, V., Pearlman, A.L. (1982). Afferent and efferent connections of the striate and extrastriate visual cortex of the normal and reeler mouse. *J. Comp. Neurol., 211,* 295-308.

[49] Steward, O., and Worley, P.F. (2001). Selective targeting of newly synthesized Arc mRNA to active synapses requires NMDA receptor activation. *Neuron, 30,* 227-240.

[50] Stosiek, C., Garaschuk, O., Holthoff, K., Konnerth, A. (2003). In vivo two-photon calcium imaging of neuronal networks. *Proc. Natl. Acad. Sci. USA., 100,* 7319-7324.

[51] Tagawa, Y., Kanold, P.O., Majdan, M., Shatz, C.J. (2005). Multiple periods of functional ocular dominance plasticity in mouse visual cortex. *Nat. Neurosci., 8,* 380-388.

[52] Thompson, J.F., Hayes, L.S., Lloyd, D.B. (1991). Modulation of firefly luciferase stability and impact on studies of gene regulation. *Gene, 103,* 171-177.

[53] Tropea, D., Van Wart, A., Sur, M. (2009). Molecular mechanisms of experience-dependent plasticity in visual cortex. *Phil. Trans. R. Soc. B, 364,* 341-355.

[54] Vazdarjanova, A., and Guzowski, J.F. (2004). Differences in hippocampal neuronal population responses to modifications of an environmental context: evidence for distinct, yet complementary, functions of CA3 and CA1 ensembles. *J. Neurosci., 24,* 6489-6496.

[55] Wagor, E., Mangini, N.J., Pearlman, A.L. (1980). Retinotopic organization of striate and extrastriate visual cortex in the mouse. *J. Comp. Neurol., 193,* 187-202.

[56] Wallace, C.S., Lyford, G.L., Worley, P.F., Steward, O. (1998). Differential intracellular sorting of immediate early gene mRNAs depends on signals in the mRNA sequence. *J. Neurosci., 18,* 26-35.

[57] Wang, K.H., Majewska, A., Schummers, J., Farley, B., Hu, C., Sur, M., Tonegawa, S. (2006). In vivo two-photon imaging reveals a role of Arc in enhancing orientation specificity in visual cortex. *Cell, 126,* 389-402.

[58] Wang, G., Cong, W., Shen, H., Qian, X., Henry, M., Wang, Y. (2008). Overview of bioluminescence tomography-a new molecular imaging modality. *Front. Biosci., 13,* 1281-1293.

[59] Wang, Q., and Burkhalter, A. (2007). Area map of mouse visual cortex. *J. Comp. Neurol., 502,* 339-357.

In: Visual Cortex: Anatomy, Functions ... ISBN: 978-1-62100-948-1
Editors: J.M. Harris et al. pp. 185-193 © 2012 Nova Science Publishers, Inc.

Chapter 7

MOLECULAR SIGNATURES OF PARALLEL PATHWAYS IN THE VISUAL THALAMUS

Hiroshi Kawasaki *

[1] Department of Molecular and Systems Neurobiology,
Graduate School of Medicine, The University of Tokyo,
Tokyo 113-0033, Japan.
[2] Global COE Program "Comprehensive Center of Education and Research for Chemical Biology of the Diseases",
The University of Tokyo, Tokyo 113-0033, Japan.

ABSTRACT

The visual thalamus conveys visual information detected by the retina to the visual cortex along parallel pathways with distinct anatomical and physiological characteristics. This group of pathways is comprised of the magnocellular, the parvocellular, and the koniocellular pathways. Although considerable progress has been made with regard to our knowledge of the anatomical circuitry and physiological properties that distinguish these three parallel pathways, our molecular understanding of the parallel pathways is still insufficient. This is, at least partially, because these pathways are not well-developed in mice, which

* Corresponding author: Hiroshi Kawasaki, MD, PhD., Department of Molecular and Systems Neurobiology, Graduate School of Medicine, The University of Tokyo, Hongo 7-3-1, Bunkyo-ku, Tokyo 113-0033, Japan. Tel: +81-3-5841-3616, Fax: +81-3-3813-2739. E-mail address: kawasaki@m.u-tokyo.ac.jp

are commonly used for molecular investigations. Recent studies of these pathways, therefore, have employed higher mammals such as primates and carnivores. In this chapter, I summarize recent findings regarding the molecular investigations of the visual thalamus, especially focusing on the parallel pathways in higher mammals. I also discuss issues to be addressed and potential future directions of this field.

INTRODUCTION

In the visual system, information detected by the retina is conveyed to the cerebral cortex via the dorsal lateral geniculate nucleus (dLGN) of the thalamus along parallel pathways comprised of neurons with distinct morphologies, connections, and neurochemical and physiological characteristics [Defoe and Van Essen, 1988; Felleman and Van Essen, 1991; Hendry and Reid, 2000; Livingstone and Hubel, 1987; Maunsell, 1992; Sherman and Guillery, 2004; Sherman and Spear, 1982]. In carnivores, the three physiological retino-geniculo-cortical parallel pathways are known as X, Y and W pathways; the X and Y pathways are comparable to the parvocellular (P) and magnocellular (M) pathways, respectively, in primates [Sherman and Guillery, 2004]. The X (or P) pathway contributes to form discrimination and color recognition, whereas the Y (or M) pathway is primarily concerned with information about object motion [Defoe and Van Essen, 1988; Livingstone and Hubel, 1987].

These distinct pathways originate in different populations of retinal ganglion cells (RGCs) and are segregated into distinct regions in the dLGN. In carnivores, the X and Y pathways project to distinct but intermingled populations of relay neurons in the A and A1 layers [Sherman and Spear, 1982]. The C layers receive input mainly from the W pathway and partially from the Y pathway. The segregation of these pathways in the dLGN is increasingly pronounced in animals with highly developed visual systems (e.g. primates) [Butler and Hoods, 2005].

Despite their significance for visual processing, the molecular correlates of these parallel pathways are not well understood. This is, at least partially, because these pathways are not well-developed in mice, which are commonly used for molecular investigations. Recent studies of these pathways, therefore, have employed higher mammals such as primates and carnivores. In this chapter, I summarize recent findings regarding the molecular investigations of the visual thalamus, especially focusing on the parallel pathways in higher mammals. I also discuss remaining issues to be addressed and potential future directions.

THE FEATURES OF THE LGN FOR UNCOVERING PATHWAY-SPECIFIC GENES

To successfully uncover pathway-specific molecules using microarray, it is important to select appropriate brain regions. When choosing brain regions, there are at least two important considerations to be taken into account. First, it is desirable to use brain regions where target neurons are accumulated spatially and separated from other neurons. For example, in the retina, although different types of RGCs exist, they are spatially intermingled, and therefore it is difficult to isolate a single type of RGCs from the retina. By contrast, X and Y cells are accumulated in the inner dLGN, and W cells are located in the outer dLGN in ferrets. Similarly, the M and P pathways are clearly segregated in the monkey dLGN: layers 1 and 2 receive the input from the M pathway, whereas layers 3-6 receive afferents from the P pathway. The second consideration is how to identify pathway-specific regions. The layers of the ferret and monkey dLGN mentioned above have clear cytoarchitectonic structures and can be easily identified in adults. Even early in development, when these cytoarchitectonic structures are not well-developed, injection of fluorescent neuronal tracers such as cholera toxin B subunit (CTB) into the eyes can clearly visualize LGN layers. These points make the dLGN an ideal brain region to explore pathway-specific molecules.

Magnocellular Enriched Genes

Earlier studies discovered that two monoclonal antibodies, Cat-301 and SMI-32, recognize M/Y cells in the adult dLGN. Cat-301, which recognizes non-glycosylated form of the extracellular matrix protein aggrecan [Matthews et al., 2002], stains perineuronal nets around M/Y cells in carnivores and primates [Hendry et al., 1984; Hockfield and Sur, 1990], though the immunoreactivity only appears late in development [Sur et al., 1988]. SMI-32 recognizes a non-phosphorylated epitope on neurofilament H (NEFH), and stains cells with Y cell-like morphology in the adult cat dLGN [Bickford et al., 1998].

To identify additional molecular correlates of these parallel pathways, our and other labs have used microarray to explore patterns of gene expression in the visual thalamus of higher mammals [Kawasaki et al., 2004; Murray et al., 2008]. Previously we fabricated a custom cDNA microarray and examined gene expression profiles in the ferret dLGN. Our microarray analyses uncovered several molecules with specific expression patterns in the ferret dLGN. Among them was PCP4, which was preferentially expressed in Y cells of the ferret dLGN [Kawasaki et al., 2004].

Another group performed genome-wide screening using the monkey dLGN and successfully demonstrated the detailed expression patterns of candidate molecules [Murray et al., 2008]. Among them were BRD4, CAV1, EEF1A2, FAM108A1, KCNA1, PPP2R2C, SFRP2, NEFL, NEFH and INα. CAV1 and SFRP2 expression was found in all principal layers, but stronger signals were observed in the M layers. BRD4, EEF1A2, FAM108A1, KCNA1, PPP2R2C, NEFL, NEFH and INα were expressed at high levels within the M layers of the dLGN. Furthermore, using these molecules, the authors investigated the potential interrelatedness of the layer-specific markers and found that the Wnt/β-catenin signaling pathway was overrepresented in this gene set.

Another study also found two molecules predominantly expressed in the monkey dLGN using microarray [Prasad et al., 2000; Prasad et al., 2002]. They dissected the M and P layers differentially and compared gene expression patterns using filter hybridization screening. After a list of candidate genes was made, they examined the expression patterns of candidate molecules using immunohistochemistry. They found that NEFM and αβ-crystallin were predominantly expressed in the M layers of the monkey LGN.

Koniocellular Enriched Genes

Previous immunohistochemical studies identified two molecules preferentially expressed in the K layers in the monkey dLGN [Hendry and Calkins, 1998; Hendry and Reid, 2000]. CaMKIIα and calbindin-D were reported to be expressed in the S and intercalated layers, which correspond to the K pathway in the monkey dLGN [Benson et al., 1991; Diamond et al., 1993; Hendry and Calkins, 1998; Hendry and Yoshioka, 1994; Johnson and Casagrande, 1995; Jones and Hendry, 1989; Yan et al., 1996]. Another study focused on the extracellular matrix proteins and found that SC1 was abundantly expressed in the K layers of the monkey dLGN [Takahata et al., 2010].

Compared with the number of M-specific genes, those preferentially expressed in the koniocellular (K) layers is relatively small. Similarly, genome-wide screening studies to look for K-specific molecules are not common. This could be because the selective isolation of the K layers is relatively difficult compared to that of other pathways. In the monkey dLGN, K cells are located between the M and P layers and form thin intercalated layers. Precise isolation of the K layers by using laser capture microdissection might facilitate the identification of K-specific molecules.

Future Directions

One of the important future directions would be the identification of P/X-specific molecules. Although several M-specific and K-specific molecules have become available as mentioned above, P-specific molecules, whose detailed expression patterns were examined by using immunohistochemistry and/or *in situ* hybridization, have not been reported. In earlier studies, to identify P-specific molecules, the parvocellular LGN layers were dissected out from monkeys and analyzed using microarray [Lachance and Chaudhuri, 2007; Murray et al., 2008; Prasad et al., 2000], and this pioneering research provided lists of P-specific candidate molecules. It was reported that TCF7L2 was expressed in the parvocellular layers in the monkey dLGN, but it was also expressed in the koniocellular layers [Murray et al., 2008]. P-specific molecules remain to be identified.

Another future direction would be the investigation of the function of the molecules expressed in the dLGN. Specification of the parallel pathways in the dLGN probably involves activity-independent molecular mechanisms. Prenatal monocular enucleation does not prevent the formation of magnocellular and parvocellular zones [Rakic, 1981], and layer-specific cytoarchitecture emerges despite early binocular enucleation [Brunso-Bechtold et al., 1983; Guillery et al., 1985]. Moreover, optic axons project selectively and directly to either magnocellular or parvocellular zone of the developing dLGN [Meissirel et al., 1997]. Therefore, some of the intriguing molecules yet to be found are pathway-specific transcription factors, which might be involved in cell-type specification early in development. Transcription factors have been reported to mediate cell-type specification in a variety of brain regions during development. Identification of such pathway-specific transcription factors would lead to a mechanistic understanding of development of the parallel pathways.

CONCLUSION

In this article, I have summarized recent findings regarding cell type-specific molecules in the dLGN of higher mammals. Molecular investigation using the dLGN of higher mammals might also have significant impacts on research using other brain regions. Compared with mice, higher mammals have several fundamental differences in their brain structures. For example, while carnivores and primates have ocular dominance columns (ODCs) in the visual cortex and gyri of the cerebral cortex, ODCs and gyri have not been found in mice. Experimental strategies and techniques used for examining molecules in the dLGN of higher mammals might be applicable to research of other brain regions such as ODCs and gyri.

Recently, transgenic techniques using higher mammals have been reported [Lois et al., 2002; Sasaki et al., 2009]. Slice cultures of the dLGN might also be useful for examining the functions of molecules expressed in the dLGN [Iwai and Kawasaki, 2009]. These techniques, combined with an understanding of the molecular organization of the visual thalamus, may open the door for exploring the molecular mechanisms underlying the development, physiological functions and pathophysiology of the visual thalamus.

ACKNOWLEDGMENTS

I apologize to numerous authors whose papers we were not able to cite due to space limitations. I am grateful to Drs. Shoji Tsuji (The University of Tokyo), Haruhiko Bito (The University of Tokyo), Takashi Kadowaki (The University of Tokyo), Makoto Araie (The University of Tokyo), Eisuke Nishida (Kyoto University), Yoshiki Sasai (RIKEN-CDB) and Shigetada Nakanishi (Osaka Bioscience Institute) for their continuous encouragement and warm support. I thank Kawasaki lab members for their help. This work was supported by Global COE Program “Comprehensive Center of Education and Research for Chemical Biology of the Diseases” from MEXT, Grant-in-Aid for Scientific Research from MEXT, Human Frontier Science Program (HFSP). This work was also supported by Takeda Science Foundation,

Astellas Foundation for Research on Metabolic Disorders, the Danone Institute of Japan, the Life Science Foundation of Japan, the Kurata Memorial Hitachi Science and Technology Foundation, the Novartis Foundation for the Promotion of Science, Mitsubishi Foundation, Fukuda Foundation for Medical Technology, Yamada Science Foundation, Hokuto Foundation, and Santen Pharmaceutical.

REFERENCES

Benson, D.L., Isackson, P.J., Hendry, S.H., and Jones, E.G. (1991). Differential gene expression for glutamic acid decarboxylase and type II calcium-calmodulin-dependent protein kinase in basal ganglia, thalamus, and hypothalamus of the monkey. *J. Neurosci. 11*, 1540-1564.

Bickford, M.E., Guido, W., and Godwin, D.W. (1998). Neurofilament proteins in Y-cells of the cat lateral geniculate nucleus: normal expression and alteration with visual deprivation. *J. Neurosci. 18*, 6549-6557.

Brunso-Bechtold, J.K., Florence, S.L., and Casagrande, V.A. (1983). The role of retinogeniculate afferents in the development of connections between visual cortex and the dorsal lateral geniculate nucleus. *Brain Res. 312*, 33-39.

Butler, A.B., and Hodos, W. (2005). Visual forebrain in amniotes. In A.B. Butler, and W. Hodos (eds.), *Comparative vertebrate neuroanatomy* (pp. 523-546). New York: Wiley.

DeYoe, E.A., and Van Essen, D.C. (1988). Concurrent processing streams in monkey visual cortex. *Trends Neurosci. 11*, 219-226.

Diamond, I.T., Fitzpatrick, D., and Schmechel, D. (1993). Calcium binding proteins distinguish large and small cells of the ventral posterior and lateral geniculate nuclei of the prosimian galago and the tree shrew (Tupaia belangeri). *Proc. Natl. Acad. Sci. USA 90*, 1425-1429.

Felleman, D.J., and Van Essen, D.C. (1991). Distributed hierarchical processing in the primate cerebral cortex. *Cereb. Cortex 1*, 1-47.

Guillery, R.W., LaMantia, A.S., Robson, J.A., and Huang, K. (1985). The influence of retinal afferents upon the development of layers in the dorsal lateral geniculate nucleus of mustelids. *J. Neurosci. 5*, 1370-1379.

Hendry, S.H., and Calkins, D.J. (1998). Neuronal chemistry and functional organization in the primate visual system. *Trends Neurosci. 21*, 344-349.

Hendry, S.H., Hockfield, S., Jones, E.G., and McKay, R. (1984). Monoclonal antibody that identifies subsets of neurones in the central visual system of monkey and cat. *Nature 307*, 267-269.

Hendry, S.H., and Reid, R.C. (2000). The koniocellular pathway in primate vision. *Annu. Rev. Neurosci.* 23, 127-153.

Hendry, S.H., and Yoshioka, T. (1994). A neurochemically distinct third channel in the macaque dorsal lateral geniculate nucleus. *Science 264*, 575-577.

Hockfield, S., and Sur, M. (1990). Monoclonal antibody Cat-301 identifies Y-cells in the dorsal lateral geniculate nucleus of the cat. *J. Comp. Neurol. 300*, 320-330.

Iwai, L., and Kawasaki, H. (2009). Molecular development of the lateral geniculate nucleus in the absence of retinal waves during the time of retinal axon eye-specific segregation. *Neuroscience 159*, 1326-1337.

Johnson, J.K., and Casagrande, V.A. (1995). Distribution of calcium-binding proteins within the parallel visual pathways of a primate (Galago crassicaudatus). *J. Comp. Neurol. 356*, 238-260.

Jones, E.G., and Hendry, S.H. (1989). Differential Calcium Binding Protein Immunoreactivity Distinguishes Classes of Relay Neurons in Monkey Thalamic Nuclei. *Eur. J. Neurosci. 1*, 222-246.

Kawasaki, H., Crowley, J.C., Livesey, F.J., and Katz, L.C. (2004). Molecular organization of the ferret visual thalamus. *J. Neurosci. 24*, 9962-9970.

Lachance, P.E., and Chaudhuri, A. (2007). Gene profiling of pooled single neuronal cell bodies from laser capture microdissected vervet monkey lateral geniculate nucleus hybridized to the Rhesus Macaque Genome Array. *Brain Res. 1185*, 33-44.

Livingstone, M.S., and Hubel, D.H. (1987). Psychophysical evidence for separate channels for the perception of form, color, movement, and depth. *J. Neurosci. 7*, 3416-3468.

Lois, C., Hong, E.J., Pease, S., Brown, E.J., and Baltimore, D. (2002). Germline transmission and tissue-specific expression of transgenes delivered by lentiviral vectors. *Science 295*, 868-872.

Matthews, R.T., Kelly, G.M., Zerillo, C.A., Gray, G., Tiemeyer, M., and Hockfield, S. (2002). Aggrecan glycoforms contribute to the molecular heterogeneity of perineuronal nets. *J. Neurosci. 22*, 7536-7547.

Maunsell, J.H. (1992). Functional visual streams. *Curr. Opin. Neurobiol. 2*, 506-510.

Meissirel, C., Wikler, K.C., Chalupa, L.M., and Rakic, P. (1997). Early divergence of magnocellular and parvocellular functional subsystems in the embryonic primate visual system. *Proc. Natl. Acad. Sci. USA 94*, 5900-5905.

Murray, K.D., Rubin, C.M., Jones, E.G., and Chalupa, L.M. (2008). Molecular correlates of laminar differences in the macaque dorsal lateral geniculate nucleus. *J. Neurosci. 28*, 12010-12022.

Prasad, S.S., Kojic, L.Z., Lee, S.S., Chaudhuri, A., Hetherington, P., and Cynader, M.S. (2000). Identification of differentially expressed genes in the visual structures of brain using high-density cDNA grids. *Mol. Brain Res. 82*, 11-24.

Prasad, S.S., Schnerch, A., Lam, D.Y., To, E., Jim, J., Kaufman, P.L., and Matsubara, J.A. (2002). Immunohistochemical investigations of neurofilament M' and alphabeta-crystallin in the magnocellular layers of the primate lateral geniculate nucleus. *Mol. Brain Res. 109*, 216-220.

Rakic, P. (1981). Development of visual centers in the primate brain depends on binocular competition before birth. *Science 214*, 928-931.

Sasaki, E., Suemizu, H., Shimada, A., Hanazawa, K., Oiwa, R., Kamioka, M., Tomioka, I., Sotomaru, Y., Hirakawa, R., Eto, T., *et al.* (2009). Generation of transgenic non-human primates with germline transmission. *Nature 459*, 523-527.

Sherman, S.M., and Guillery, R.W. (2004). Thalamus. In G.M. Shepherd (ed.), *The synaptic organization of the brain* (pp. 311-359). New York: Oxford University Press.

Sherman, S.M., and Spear, P.D. (1982). Organization of visual pathways in normal and visually deprived cats. *Physiol. Rev. 62*, 738-855.

Sur, M., Frost, D.O., and Hockfield, S. (1988). Expression of a surface-associated antigen on Y-cells in the cat lateral geniculate nucleus is regulated by visual experience. *J. Neurosci. 8*, 874-882.

Takahata, T., Hashikawa, T., Tochitani, S., and Yamamori, T. (2010). Differential expression patterns of OCC1-related, extracellular matrix proteins in the lateral geniculate nucleus of macaque monkeys. *J. Chem. Neuroanat. 40*, 112-122.

Yan, Y.H., Winarto, A., Mansjoer, I., and Hendrickson, A. (1996). Parvalbumin, calbindin, and calretinin mark distinct pathways during development of monkey dorsal lateral geniculate nucleus. *J. Neurobiol. 31*, 189-209.

INDEX

C

D

E

F

G

H

I

J

K

L

M

N

O

P

R

S

T

U

V

W

Y